Scrum Narrative and PSM™ I

Compliant with the latest version of The Scru

Mohammed Musthafa Soukat

ISBN: 9781521475461

First Published Nov 24, 2015

Rev. 3.5

2021 Edition, Revised Mar 19, 2021

Editor in Chief: **Samantha Mason**

Get a free copy of the latest edition: Thank you for choosing this book. For our records, please drop an email to psm1examguide@gmail.com with a copy of your purchase receipt. Any revised edition within six months after your purchase will be sent to your email address.

Also by Mohammed Musthafa Soukath Ali. Available at leading online retailers:

1. The Road to Agile Coaching: A Life Changing Leadership Role
2. Get SAFe Now: A Lightning Introduction to the Most Popular Scaling Framework on Agile
3. Guide to Pass PSPO 1 Certification from Scrum.org

#

Acknowledgement

Perfection by an Individual is by accident. Perfection by a Team is by design. Thank you TCS team for all that you did.
#

Table of Contents

#

Preface

In the last decade, I witnessed many successful Scrum transformations in the software industry. I coached many teams that served customers in different segments and saw the way these customers became convinced of the benefits of this new way of working.

Scrum continues to be an effective way of working. This new way is rapidly making its way into many organizations, in particular the software development divisions, due to the transformational results it delivers. So, there is increasing need for today's professionals to understand the transformation Scrum brings to their way of working. The authentic source to know about Scrum and the approved body of knowledge on Scrum is "The Scrum Guide" authored by Jeff Sutherland and Ken Schwaber. The sixteen-page Scrum Guide is packed with rich, insightful content. Yet for new entrants or those with minimal exposure to Scrum, the guide is too dense to absorb.

A storybook for Executives

The first part of this book narrates Scrum in a lightweight and engaging fashion. It is an executive summary for organization leaders. While many books on Scrum position it as a collection of roles and practices, this book introduces Scrum's transformational roots that bring in a newer way of working with far reaching effects.

A coaching book for Professionals getting into Scrum

The second part delves deeper into the finer aspects of Scrum, which is required for professionals working on the ground. It walks through the stages of the Scrum Journey from Starting, Executing, to Closing. Unlike many books that are limited to the mechanics of Scrum, this book addresses plenty of practical questions including business management, team and task management, product engineering, etc. This understanding goes beyond the Scrum theory and enhances job skills.

An all-in-one guidebook for PSM 1 Certification

This entire book is an all-in-one guidebook for PSM 1 assessment preparation. PSM 1 is a high-quality Scrum certificate administered by Scrum.org that is guided by Ken Schwaber, one of the original co-authors of Scrum. Unlike many other Scrum certificates, PSM is not a vanity means to claim Scrum knowledge, but a rigorous assessment of the knowledge in the original Scrum. Though there are multiple scattered materials available, there is no comprehensive guide for PSM assessment preparation. This book is a one-stop source including guidance on understanding Scrum, registration, preparation, and extensive practice. The book augments the Scrum narrative with exam preparation tips, quick tests, and a full-blown assessment mimicking the real assessment. It provides 250+ PSM 1 assessment-related questions to practice.

What is unique about this book?

Sticks to the authentic version of Scrum: There are many sources available that teach Scrum and provide training material for the PSM. It is common for them to provide a muddied version of Scrum associating it with activities, artifacts, and sub-techniques that are not prescribed or endorsed by the Scrum framework. Some of them even misinterpret Scrum fundamentals. Such sources may add confusion and cloud your attempt to understand Scrum. Also, such an understanding will make one choose incorrect answers in the PSM assessment. This book articulates the original, unpolluted Scrum framework as defined in its authentic source "The Scrum Guide." Wherever there are exceptions or additional pointers, they are highlighted by a DE-TOUR tag.

Anchors the learning using an Active Learning technique: This book helps the reader to absorb the deeper meaning behind Scrum by means of Active Learning. Active Learning used in this book refers to not just passive reading of the content, but taking frequent pauses to answer questions about what was just read, and then thinking, analyzing, and inferring the meaning through answering those questions. These granular interpretations are usually not evident in normal reading, but pop out when backed by questions around them.

--------------------------Disclaimer--------------------------
The author or any organization the author represents at the time of writing this book, is NOT a certifying body for any Scrum.org certifications (PSM I, PSPO I, etc.) and has no role or influence in granting any of the Scrum.org certifications. This book is meant to help individuals prepare for the PSM 1 assessment.
-------------------------- Disclaimer --------------------------

#

Why PSM 1?

- The PSM 1 exam is administered by a company guided by Ken Schwaber, who is one of the co-founders of Scrum.
- PSM is the most meritorious certificate of all that certify Scrum. Other certifications can easily be obtained by attending a class or through an exam where failing could only happen by intention.
- PSM assessment requires good knowledge of the original version of Scrum, and its passing requirements are high (85%). Such stringent criteria provide more teeth to its certificate than the other certifications provided in the marketplace.
- PSM does not require any mandatory training courses.
- PSM assesses the knowledge on authentic Scrum.
- Nominal fee of $150.
- The assessment can be taken from anywhere with a computer and internet connection.
- Once acquired the certificate does not need renewing.

#

Part 1- Executive narration of Scrum

The best way to predict the future is - to create the near future and listen to the feedback.

#

Chapter 1 - A new approach for complex problems

--------------------------DE-TOUR--------------------------

The Scrum Guide directly introduces Scrum as a framework for developing and sustaining complex products. However, it is important to know the old ways of developing products in order to better understand the contrast that Scrum brings. In this chapter, the following are included for clarity and context. They are not part of The Scrum Guide:
Waterfall, Traditional project plan, Late feedback, Cone of uncertainty, Scientific approaches

--------------------------DE-TOUR--------------------------

How a Product is built using the traditional way – Waterfall

Organizations create strategies for business purposes. Some of these strategies aim at building product capabilities. For example, a software product building strategy may include a Customer Relationship Management System, Billing and Payments, Mobile Channel, New Product Introduction, etc.

Many organizations use a process methodology called waterfall in their projects to build products. Two types of people are needed to define the business and development aspects of project plans.

<u>The businesspeople</u>: They define what the product should do. To define the product needs, they make long-term predictions about the future. Some examples of predictions include what the value of the product will be, how it will be received in the market, and more importantly what that market will be when the product is actually released. Predictions are based on many assumptions and projections built on what is known today, which may be years before the product is actually released.

<u>The project managers or planners</u>: They plan how to build the product. Based on the product needs deduced from the business predictions, they come up with a sequence of activities like Analyzing, Designing, and so on. They also predict (estimate) the future cost and time of these activities to find the total project cost and time. Such estimations are based on many assumptions and future projections of the product needs, people competency and behavior, project technology, etc.

Why the name waterfall?

The outcome from the waterfall-based project planning is a detailed plan.

Later, during the execution of this plan, future changes may invalidate the current plan. Such changes may require re-planning project activities, sometimes from the basic product definition, and re-doing all the activities. Such revisiting, re-planning, re-doing, is called Iterating. In waterfall, iteration is seen as the result of bad planning. So, when people in waterfall encounter a change, they reactively go through an elaborate change control process to resist the change in order to keep the original plan intact.

A waterfall-based project plan expects that the sequence of steps in the plan will go forward without requiring changes. They resist changes when they do occur. So, the plan is designed such that the activities only go forward without iterating back again, and hence the name waterfall. It is just a waterfall, the water falls only forward, not backward.

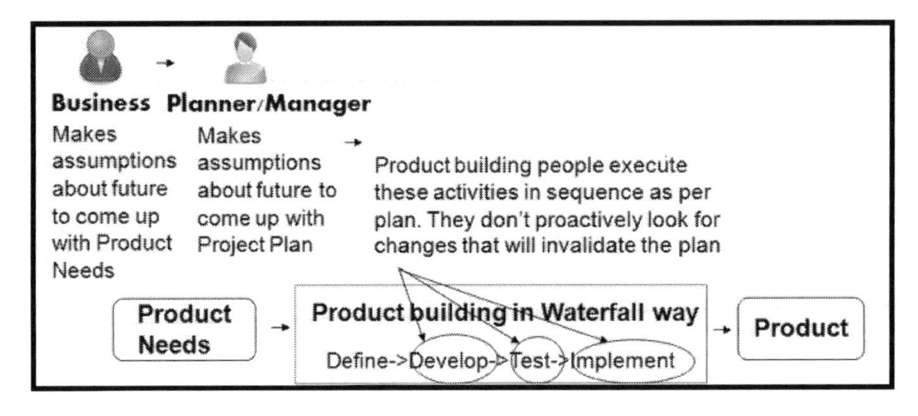

Fig. 1- Waterfall way of building products

-------------------Question- 1-------------------------

In the waterfall methodology, the duration of the activities, like development, testing, etc. are

a) Planned to be fixed irrespective of any calculation (time-boxed).
b) Predicted using some calculations based on today's understanding and assumptions.
c) Never decided upfront.

-------Answer-------

The waterfall activities are estimated and calculated well in advance based on today's knowledge and assumptions. Their duration will vary depending on the calculations. They are not time-boxed; they do not have fixed durations. Correct answer is 'b.'

-------------------Question- 1-------------------------

Waterfall – The tale of late feedback

After executing the complete plan, usually after a long period of time, the product is delivered to the business/users in a big one-time release. These people see the outcome and provide feedback about the product. The business may choose to test the market and release the product. So, the market also may provide feedback. If the feedback is positive and indicates the larger acceptance of the product, everyone is happy.

However, this is not always the case. The assumptions made by the business about user behavior may be invalid. The interpretation of the Project Managers/Planners about what the businesspeople wanted may be incorrect. Also, the external factors, market receptiveness, and assumptions might have changed.

Sometimes, the feedback may be about a new insight that requires major modification to the product or the product itself is identified as obsolete. Such feedback is called Late Feedback since the waterfall project may already be closed or it may involve too large an effort to incorporate the feedback so late.

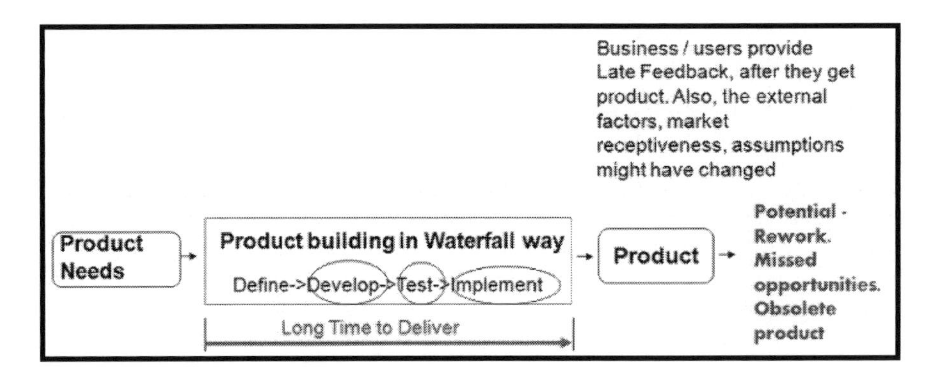

Fig. 2- Late Feedback in Waterfall

-------------------Question- 2-------------------------

In waterfall, the project team often gets early feedback from customers or markets about the product they are building, so they can act on this feedback on time.

a) True

b) False

-------Answer-------

Waterfall-based projects rarely offer any opportunity to get early feedback about the product. Correct answer is 'b.'

-------------------Question- 2-------------------------

Complex problems

Product building is a complex problem. Most often it is not just an isolated product development but may involve integration of the product into the larger organizational system.

For example, in software development as technology evolves and markets change, organizations need to continuously adopt the newer software developments and enhancements into their existing complex web of technology infrastructure.

This complexity is further multiplied by the presence of many other factors such as different people, processes, technology components, and so on.

Complexity becomes more dynamic – Complex Adaptive Problems

Product building is not only complex. Multiple factors involved in building complex products also vary over time. For example, in a software project a software component that worked in a small-scale system, which has limited users today, may not work when the system becomes huge with large numbers of users tomorrow. Similarly, some of the developers that worked today may not be available for the project tomorrow, and the productivity may vary.

Such time-dependent complex problems are also called complex adaptive problems.

In these problems, the amount of unknown is huge in the initial stages and will likely trend down over the course of product development. This trend is called the cone of uncertainty. Future change is certain in complex adaptive problems.

-------------------Question- 3-------------------------

What factors will increase complexity? Select all that apply.

a) Larger number of people on the project

b) Longer duration of the project

c) Batching of a large amount of features into one big release

-------Answer-------

All these factors – more people, longer duration, and only one big release – increase the complexity. Correct answers are 'a,' 'b,' and 'c.'

Note about multiple correct choices: In the actual PSM 1 assessment, if you need to choose more than one correct choice, each choice is provided as a check box (instead of a radio button). Additionally, the question will ask you to select two / three correct choices. In this book, we have used 'Select all that apply' to indicate there are multiple correct choices. This is intentional so that you can learn more by thinking through each choice.

-------------------Question- 3-------------------------

Is waterfall-based Project Planning a scientific approach?

At the initial stage of a project, planners use techniques such as Critical Path Method (CPM) to scientifically calculate the duration of the project. These calculations need some inputs such as product definition, people productivity, etc. The planners make assumptions to arrive at these inputs.

Since the project is at the initial stage, the cone of uncertainty is high and hence the assumptions have high probability of becoming incorrect.

So, if these assumptions are incorrect, output from the scientific calculation will also be incorrect. Therefore, project planning in the waterfall method is based on scientific calculation, but it does not mean that this planning is foolproof and risk-proof.

Are scientific approaches good for complex adaptive problems?

Scientific calculations and predictions are helpful when the problems are deterministic, stochastic, etc. In these types of problems, the future behavior or result can be modelled or predicted.

Complex adaptive problems are not deterministic. Deterministic requires the elements of the problem to be either constant or follow a definite mathematical model. The elements of complex adaptive problems are not predictable and do not follow a definite mathematical model.

Complex adaptive problems are also more complicated than being stochastic. While the elements of stochastic problems are random, the range or the boundaries of variation can be predicted. This boundary can be derived from data of the past. An example is the toss of a coin. Though the result of the coin toss is random, it is always within the boundary of either head or tail. The elements of complex adaptive problems are not only random, but their random behavior cannot be modeled based on the past.

In complex adaptive environments, what will happen is unknown and hard to predict. Predictions based on scientific approach may need extensive resources and recursive fine-tuning to arrive at an optimal plan. For planning one project, such a huge investment in scientific modeling is usually not justified.

-------------------Question- 4-------------------------

Building complex products like software is a complex adaptive problem. Complex adaptive problems are

a) Deterministic.

b) Stochastic.

c) Hard to predict even using past history.

-------Answer-------

Complex adaptive problems are hard to predict. Correct answer is 'c.'

-------------------Question- 4-------------------------

The New Way - Empiricism – Observation rather than prediction

Empiricism theory is based on the concept that complex problems are hard to predict.

Empiricism helps people to navigate uncertain complex problems. It requires them to take one step at a time, such as performing a small amount of work to gather experience.

1. It divides complex problems into a small scope of short-term iterations. For each iteration, just enough work is planned that can be completed within that short iteration of a few weeks. There is no big scope of work spanning a long duration.

2. It uses a small team to perform these few weeks of work. It requires them to increase the visibility of work or product information as they travel this iteration.

3. The small team creates a product Increment which is a usable outcome, at the end of these few weeks. The Increment is shared with stakeholders for inspection and feedback is solicited. The team gathers experience from this work. Stakeholders are those who have specific interest and knowledge in the product built by the team.

4. From this feedback and experience, new clarity emerges, and knowledge is obtained. The scope and plan for the next iteration is adjusted based on this new knowledge.

You can equate the traditional "trial and error" model to empiricism. Each iteration is a trial to solve a problem and gain more clarity. Based on the trial outcome and the newly found clarity, the next iteration is planned. Since each iteration is planned based on the newly found clarity from the previous iteration, the risk of unknowns is gradually reduced over time.

Reducing risk increases the probability of meeting the goal. In other words, each iteration brings the product closer to the overall Product Goal.

So, empiricism applies an iterative, incremental approach to optimize predictability and control risk.

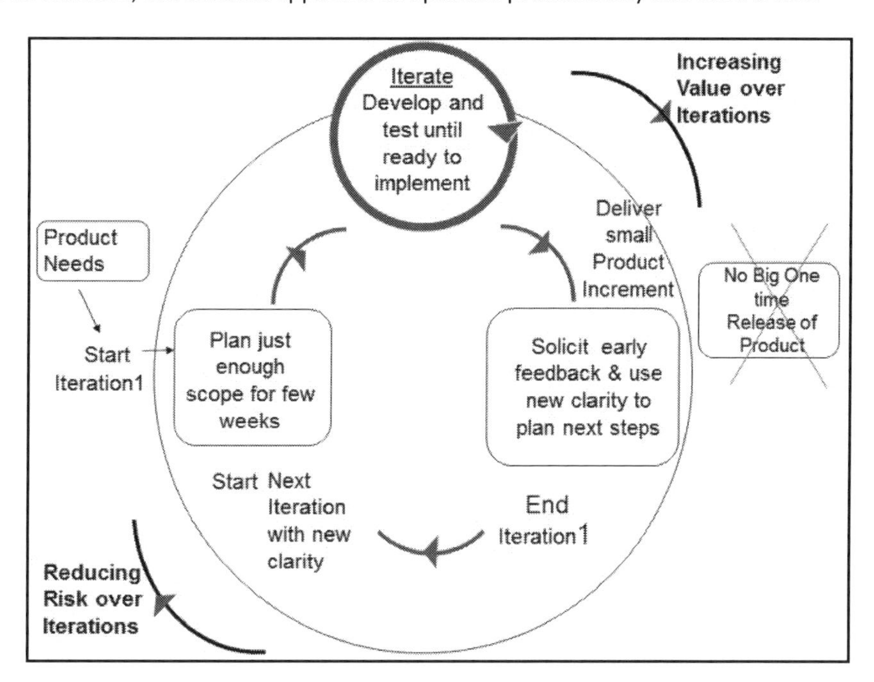

Fig. 3- Empiricism through iterative incremental approach

------------------Question- 5-------------------------

Select all that apply. Empiricism provides

a) Frequent feedback opportunities to obtain information that can be used to plan better and completely eliminate the uncertainty.

b) Frequent opportunities to discuss different possibilities.

c) Frequent opportunities to make informed decisions thus reducing risk.

-------Answer-------

Empiricism is an alternative to waterfall to manage complexity and uncertainty. In waterfall, the risk of uncertainty accumulates over long cycles. The risk is reduced by empiricism because it provides frequent feedback and course correction points. These points are where more information may be available to view different possibilities and make informed decisions.

However, empiricism does not completely eliminate uncertainty. Correct answers are 'b' and 'c.'

-------------------Question- 5-------------------------

Scrum – A new way to address complex problems

Scrum is a newer way or framework within which people can address complex adaptive problems. Scrum is founded on empiricism control theory. Scrum provides a new set of terminology to define the framework.

-------------------Question- 6-------------------------

Select the best answer. Scrum is a newer way of doing things to address complex problems. It is a newer way because

a) It offers new terminology for traditional practices.
b) It is easier to master (implement) than the traditional way.
c) It increases the opportunity to control risk and optimizes the predictability of progress.
d) It is closely associated with emerging technologies.

-------Answer-------

Scrum does introduce new terminology, but it is not the primary difference. Scrum is easy to learn, but difficult to master. There is incidental association of a lot of emerging technologies executed in Scrum, but that is not the reason for its identity as a newer way.

Scrum does not guarantee success, but it increases the likelihood of success by controlling the risks and optimizing predictability. Correct answer is 'c.'

-------------------Question- 6-------------------------

Is Scrum a proven approach?

Ken and Jeff authored Scrum two decades ago. Several organizations and practitioners have applied it with profound results. Many case studies are documented. Today, Scrum is the most widely adopted framework among all the frameworks intended to bring agility into producing software. Agility indicates an ability to respond in time to the emerging product requirements, and the term 'Agile' is an associated philosophy of software development. It is described more in Chapter 3.

Scrum events that offer opportunity for early feedback

There is a total of five events within the Scrum framework. Other than the container event Sprint, each event implements the theory of empiricism by offering an opportunity to get early feedback and the opportunity to best utilize that feedback.

Five events of Scrum

1. Sprint: Sprint is the heart of Scrum. It is like a mini project that contains the other four events.

2. Sprint Planning: This is the first event of the Sprint. A small amount of work is chosen and a plan of how to deliver that work is put together.

3. Daily Scrum: The short and quick daily recurring event within the Sprint where the team members synchronize their progress with each other and confirm the next 24-hour plan.

4. Sprint Review: This one-time event within the Sprint is where the team makes their progress visible to the stakeholders. Both the team and stakeholders collaborate to adjust their next steps.

5. Sprint Retrospective: This is the final event of the Sprint. This is an inspection and adaptation opportunity for the Scrum Team to inspect their way of working and identify potential improvements.

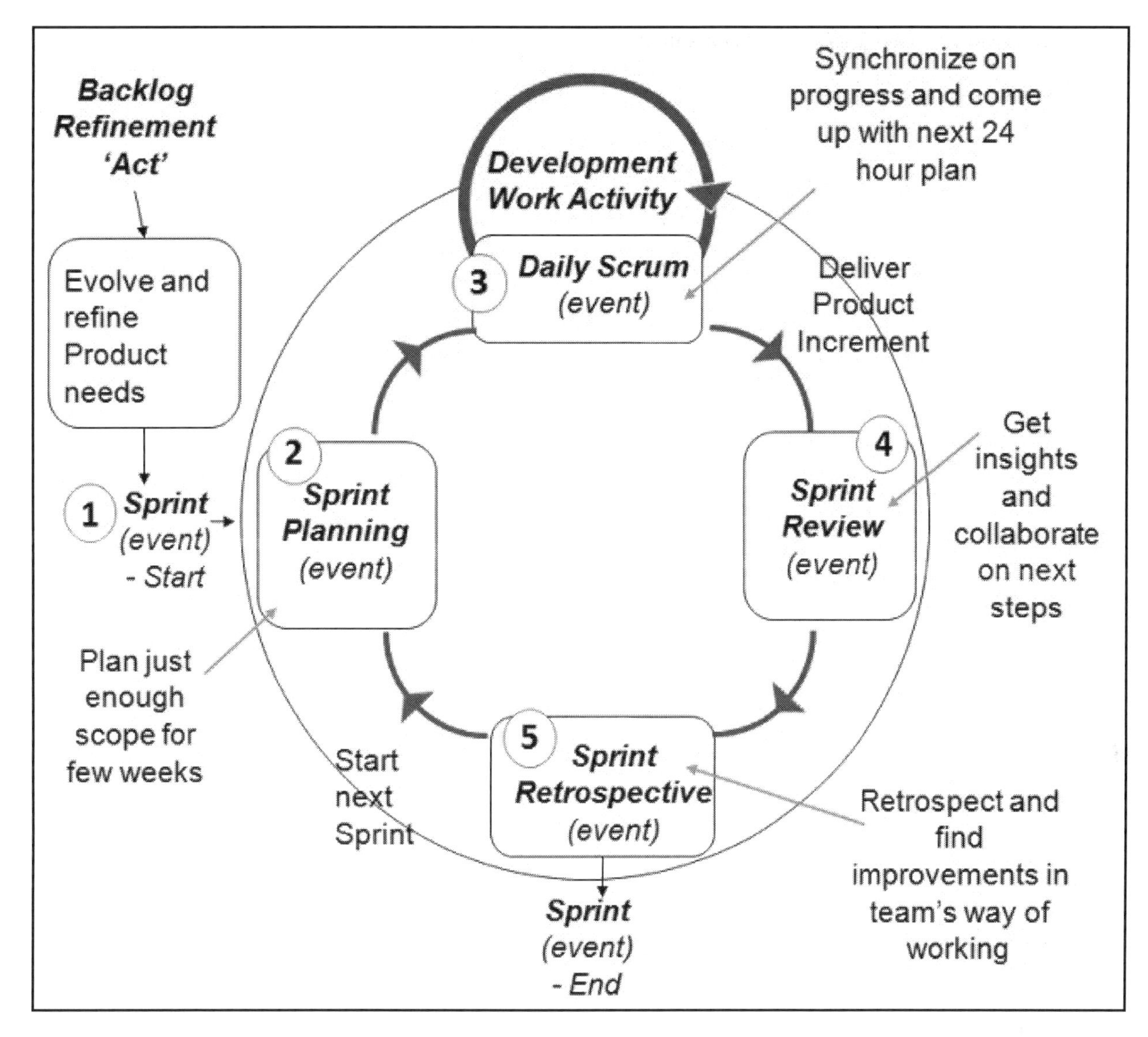

Fig. 4- Scrum events

-------------------Question- 7-a--------------------------

Scrum effectively limits the risk of unknowingly doing something wrong by

a) Having a detailed and extensive risk management plan.
b) Having multiple checkpoints and a thorough review by senior management at those checkpoints.
c) Using short iterations called Sprints.

-------Answer-------

Near the end of each Sprint, the team will get feedback on their product demonstration from the stakeholders and users to understand if their work was worthwhile. If the work outcome is identified as a waste, the cost spent is limited only to that of the weeks incurred in the Sprint. This feedback is a control and can be used to plan the next Sprint. In other words, the risk of pursuing a wrong direction is limited to the cost of one Sprint. Correct answer is 'c.'

-------------------Question- 7-a--------------------------

Scrum is increasingly used beyond product development

Though Scrum is used for complex product development and maintenance, this simple framework also found its way into many other things. It is extensively used in technology industry to build and maintain software. Scrum is even applied as a framework for teaching in schools, improving the activities in home, planning events, etc.

In many organizations, Scrum has gone beyond the software development into other functions such as marketing. Some organizations apply Scrum for overall management. Following are some of the reasons for widespread adoption of Scrum:

a) Small Scrum Team of professionals that is self-managing and highly adaptable unlike the traditional big teams that are externally managed.

b) Scrum Team is focused on one objective at a time, the Product Goal. Scrum Team has a sense of ownership towards its goal. The sense of ownership pushes the Scrum Teams to reach each other through informal networks and get things done.

c) Technology improvements such as collaboration platforms and distributed development and release environments, many of them enabled by cloud, accelerate the work though this network of teams so they can develop, release, operate and sustain the work and work products of thousands of people.

-------------------Question- 7-b-------------------------

Scrum has been used to

a) Develop products, product enhancements, and operational environments for products use.
b) Research viable markets and new product capabilities.
c) Manage the operation of organizations.
d) All of the above.

-------Answer-------

Scrum has been used to develop software, hardware, embedded software, networks of interacting function, autonomous vehicles, schools, government, marketing, managing the operation of organizations and almost everything we use in our daily lives, as individuals and societies. When the words "develop" and "development" are used in the Scrum, they refer to complex work, such as those types identified above. Correct answer is 'd.'

-------------------Question- 7-b-------------------------

-------------------Question- 7-c-------------------------

Scrum Team is focused on

a) The Product Goal.
b) Testing of the Product.
c) Design of the Product.

-------Answer-------

While the individual team members may focus on finer aspects of product building such as testing, the Scrum Team as a cohesive unit of professionals is focused on the Product Goal. Correct answer is 'a.'

-------------------Question- 7-c-------------------------

Summary

- The waterfall way of product building leads to a plan that tries to predict the future.
- The waterfall plan does not proactively look for early feedback. Feedback happens late.
- In complex adaptive problems, the future is uncertain and hard to predict. The waterfall way of predicting the future is risky.
- Empiricism advocates observation rather than prediction to navigate complex adaptive problems.
- It asserts that knowledge comes from experience and using that knowledge to make decisions based on what is known.
- Scrum is founded on empiricism that facilitates early feedback.
- Scrum has been used to manage complex product development since the 1990s.
- Scrum is a lightweight framework that helps people, teams and organizations generate value through adaptive solutions for complex problems.
- Scrum employs an iterative, incremental approach to optimize predictability and to control risk.
- The Scrum framework consists of Scrum Teams and their associated roles, events, artifacts, and rules.
- Other than the Sprint itself, which is a container for all other events, each event in Scrum is a formal opportunity to inspect and adapt.
- Scrum was initially developed for managing and developing products. Scrum is now widely used for many other things including products, services, and the management of the parent organization.

Quiz

1. Scrum is

a) A framework to generate value through adaptive solutions for complex problems.

b) A collection of the industry's best practices in information technology.

c) A body of knowledge on software engineering to build products.

-------Answer-------

Scrum is a framework based on empiricism process control theory within which people can generate value through adaptive solutions for complex problems. Correct answer is 'a.'

2. To reduce the likelihood of not meeting big commitments, Scrum uses

a) Timeboxing so the planned events can happen on time.

b) Timeboxing so the commitments will have additional buffer of time.

c) Timeboxing so the events cannot exceed a predetermined amount of time.

-------Answer-------

Scrum controls the risks associated with long term planning and big commitments by constraining the product development into shorter iterations called Sprints. Each Sprint is strictly time-boxed so they expire on the predetermined date no matter what. By timeboxing, the risk of pursuing a wrong direction is limited to the cost of one Sprint. Correct answer is 'c.'

3. Select all that apply. The formal opportunities for inspection and adaptation are

a) The Sprint.

b) The Scrum Team.

c) The Product Increment.

d) None of the above.

-------Answer-------

Four Scrum events are formal events for inspection and adaptation. They are Sprint Planning, Daily Scrum, Review, and Retrospective. Correct answer is 'd.'

4. Select all that apply. The Scrum framework consists of

a) Scrum Standards.

b) Scrum Teams.

c) Product Development Processes.

d) Roles, events, artifacts, and rules associated with Scrum Teams.

-------Answer-------

The Scrum framework consists of Scrum Teams and their associated roles, events, artifacts, and rules. Though there are standards such as Definition of Done, they are not formally called Scrum Standards. Correct answers are 'b' and 'd.'

5. As per empiricism, knowledge is acquired by

a) Experience.

b) Analytical tools.

c) Formal teaching.

-------Answer-------

Empiricism asserts that knowledge comes from experience in the past and the practice of making decisions based on what is known. Correct answer is 'a.'

#

Chapter 2 – Vehicle of Scrum - An absolute and less complex team

-------------------------DE-TOUR-------------------------

In this chapter, the following are included for clarity and context. They are not part of The Scrum Guide:
Project Manager, Existing Project Managers, Managing people aspect of the Scrum Team, Business Manager

-------------------------DE-TOUR-------------------------

Along with empiricism Scrum aims to maximize the value of people working together. The vehicle of Scrum is the Scrum Team. It is a small team that has a clear focus on product ownership and has less complexity in the way it works.

To enable this ownership and reduced complexity, Scrum lays down two strict rules:

1. Reduce management-communication and role overhead by being small and self-managing: The Scrum Team is small enough to remain nimble and large enough to complete significant work within a Sprint, typically 10 or fewer people. A Scrum Team plans, executes, and controls its own work without any one individual managing their work. The team contains only three accountabilities, Product Owner, Scrum Master, and Developer. A team that manages its own work (internally decides who does what, when, and how) is called self-managing.

2. Take full ownership by being cross-functional: Scrum Teams are cross-functional, meaning the members have all the skills necessary to create value each Sprint. A team that has all the required skills to build the product is called cross-functional.

------------------Question- 8-------------------------

What are the roles in a Scrum Team? Select all that apply.

a) Project Manager
b) Programmer
c) Tester
d) Business Analyst
e) Architect
f) Operations Analyst
g) None of the above

-------Answer-------

There are only three roles in a Scrum Team. Correct answer is 'g.'

------------------Question- 8-------------------------

Developers - create any aspect of a usable product feature each Sprint

Let's start with understanding the Developers. Fig. 5 shows the Developers of a Scrum Team. Developers perform all the development work required to convert the business needs into a useable product feature. Every one of them is called 'Developer' irrespective of their primary skill set.

For example, in a Scrum Team that builds software, each of the Developers could be a specialist in an individual area. But there is no special role assigned to them. As needed, the Developers can take up any activities such as user interface design, coding, testing, integrating, user manual creation, etc. to reach the goal. This arrangement is to reduce the role overhead and people working alone.

Note: Until 'The Scrum Guide 2020,' the developers were structured as Development Team within the Scrum Team. Starting with 'The Scrum Guide 2020,' this separate team structure is eliminated. The goal was to eliminate the concept of a separate team within a team that has led to "proxy" or "us and them" behavior between the Product Owner and Development Team. There is now just one Scrum Team focused on the same objective, with three different sets of accountabilities: Product Owner, Scrum Master, and Developers.

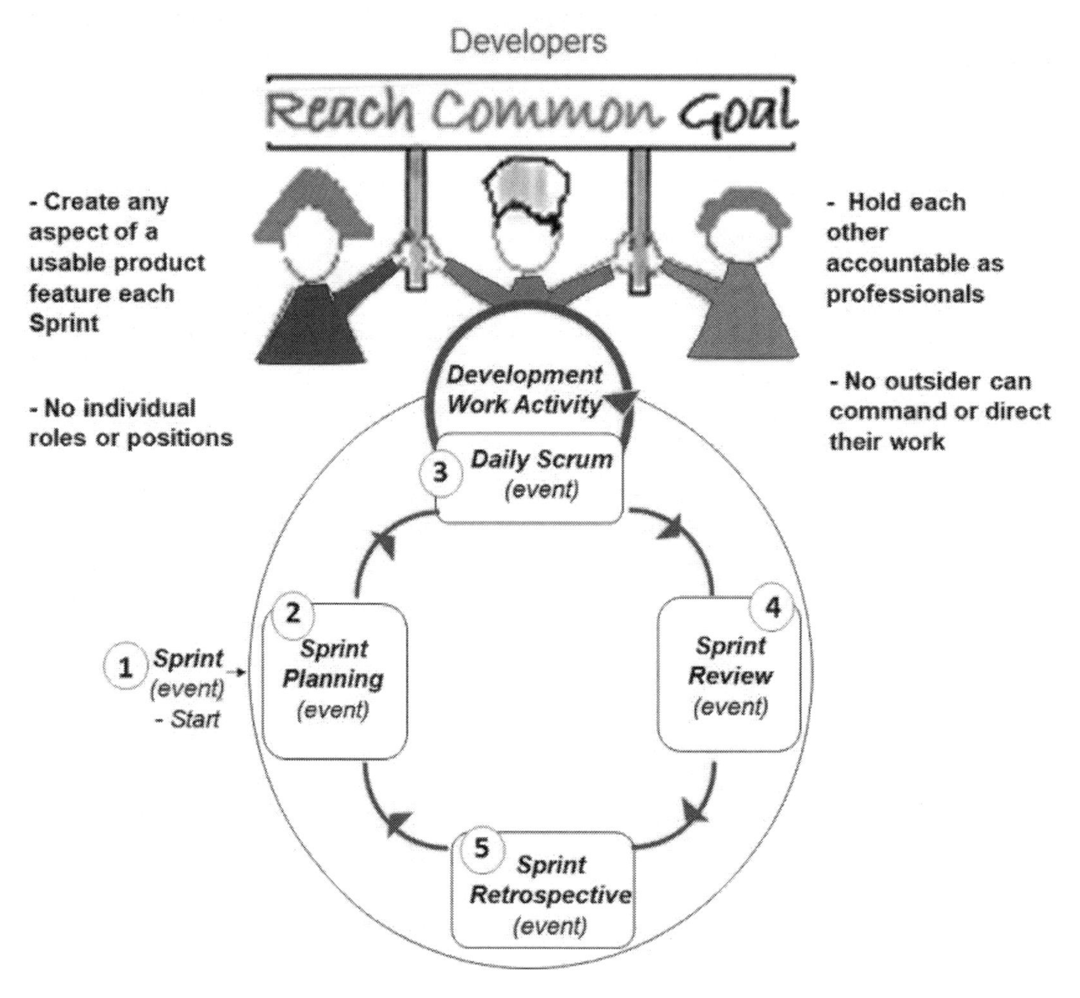

Fig. 5- Scrum- Developers

Since everyone is a Developer without a title of specialization, can the team members pursue specializations anymore?

However, while the specialists should identify themselves as part of the team and learn additional skills to collectively deliver the work, there is no barrier to personally enhancing their special competencies and continuing to specialize.

Product Owner – maximizes the value of the product resulting from the work of the Scrum Team

In waterfall, batching and big bang releases are common.

Accumulating developed product features over a long time without releasing them for use is called batching. Usually they are released together as a batch in a big-sized release (big bang fashion) in one or two limited production releases.

By batching the features for a combined release, the waterfall plan treats all these features as having the same value with no sensitivity to time.

The truth may be that different features within a batch may have different business values. So, it will make significant business sense to rank (order) the features by their business values and start delivering those more valuable features earlier.

By having the team develop features with the highest business value first, we earn more return on the team's work. Early developed features can be released to production if it makes business sense.

There is an exclusive businessperson in the Scrum Team that is responsible for keeping the product features ordered and communicating the overall Product Goal of these features together. This person is the Product Owner.

The Product Owner communicates the overall Product Goal and feeds the list of ordered product features to the team. By ensuring that the team works on higher value features first, and constantly working with them to keep them focused on the Product Goal, clarifying the business questions, they can optimize the team's work against value.

The Product Owner can choose to deliver the completed features to production often instead of batching them. By using the feedback obtained from the production usage and any market changes, the Product Owner can adjust the product features and their order to maximize the business value and control risk.

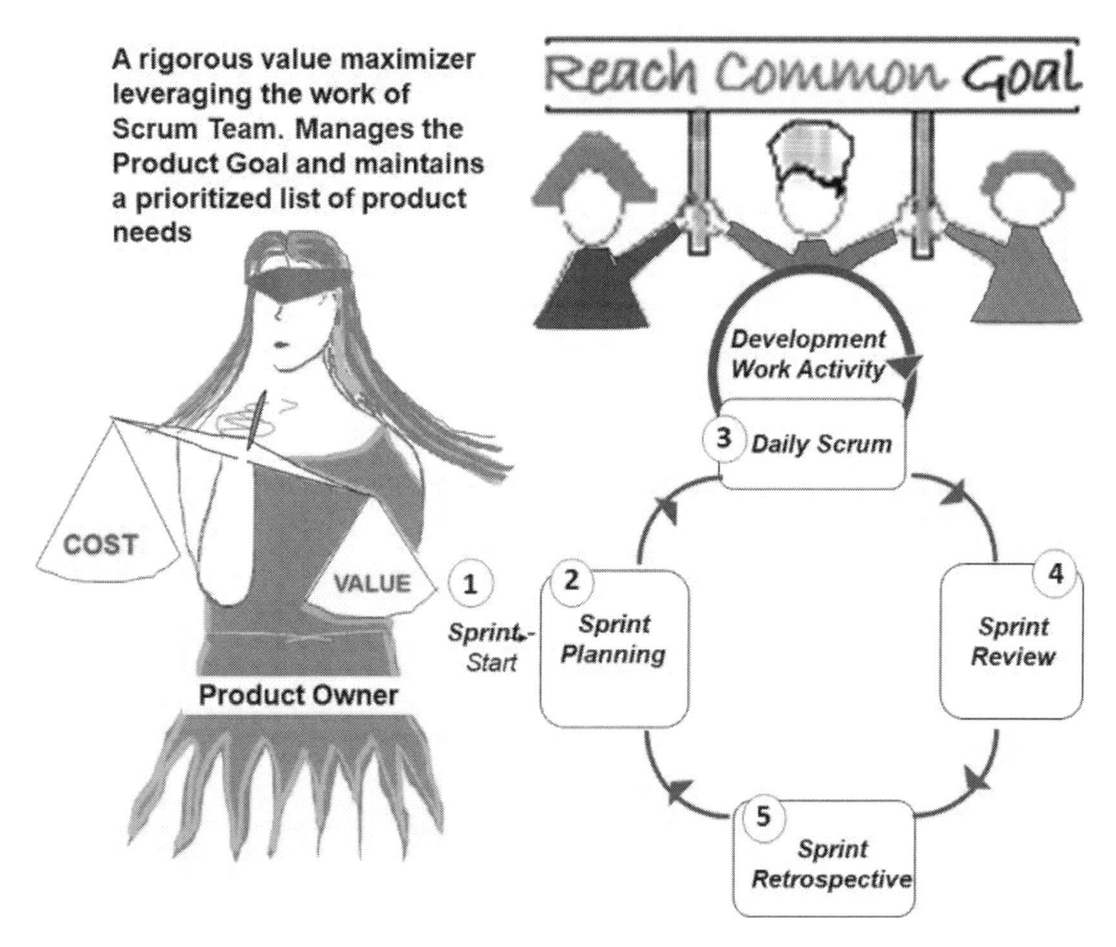

Fig. 6- Product Owner

We do have business managers even today. What is so different with the Product Owner?

The Product Owner does not resort to a traditional working style where they define all the product needs upfront based on today's knowledge, and then toss a big requirements document over the "fence" to another team for development.

The Product Owner is much more than a traditional business manager in two aspects:

1. Continuous engagement with the team: The Product Owner respects the fact that the future cannot be predicted. They adopt empiricism to continually gain clarity and refine the product. They are on the constant look out for every opportunity to maximize the value of the product. They continually work with the team to help them understand the product needs and get the best value out of their work.

2. Ultimate authority of the team's scope of work: The Product Owner leverages the dedicated Developers to continually shape the product and uncover new knowledge. So, it is essential that the team should only be working on their product needs. The team should not be "switching the context" of their work due to any external authorities directing them to do "some other" work. So, the Product Owner is the ultimate authority on what the team should work on next. Even the CEO of the organization cannot request the team to work on something else. Anyone wanting to add something to the product must go through the Product Owner.

-------------------Question- 9-------------------------

Select all that apply. In Scrum, the Product Owner who plays the business role,

a) Hands over the Product Backlog to the Developers and leaves them alone. They only meet up again during final product delivery.

b) Freezes the Product Backlog and tries not to change it.

c) Works only with designated Developers and not all the Developers.

d) Continuously collaborates with the Developers, sometimes almost every day.

-------Answer-------

The Product Owner continuously evolves the ordered list of everything that may be needed to meet the Product Goal. This list is called the Product Backlog. Since this list evolves based on frequent new insights, it requires the Product Owner to continuously work with the team to communicate these changing needs and to clarify questions about ongoing work. Correct answer is 'd.'

-------------------Question- 9-------------------------

Scrum Master – Serves the team and establishes Scrum as defined in the Scrum Guide

In Scrum, empiricism, lean thinking, and self-management drive the product development approach. A leadership role is introduced to teach and coach people about these concepts and other elements of Scrum as defined in the Scrum Guide.

This leadership model is called servant leadership. The servant-leader does not take the lead in planning or controlling the development work. Instead, the servant-leader mentors the team to manage their work themselves within the Scrum framework.

-------------------Question- 10-------------------------

In Scrum, the 'servant-leader' is the new name for the traditional role called 'Project Manager.'

a) True

b) False

-------Answer-------

The servant-leader of the self-managing team manages the implementation of principles like Self-Management, Theory of Empiricism, Lean Thinking, etc. by teaching and coaching the team. The servant-leader is neither a Project Manager nor a People Manager. Correct answer is 'b.'

-------------------Question- 10-------------------------

In Scrum, the person playing the servant-leader role is called the Scrum Master. The Scrum Master serves the team by coaching them to work together for a common goal irrespective of their individual skills. The Scrum Master mentors the team so that it becomes self-sufficient in their product development ownership. Such a self-sufficient team will frequently create a working product Increment, get early feedback in order to re-plan based on emerging insights, and solve the problems by their collective wisdom and collaboration.

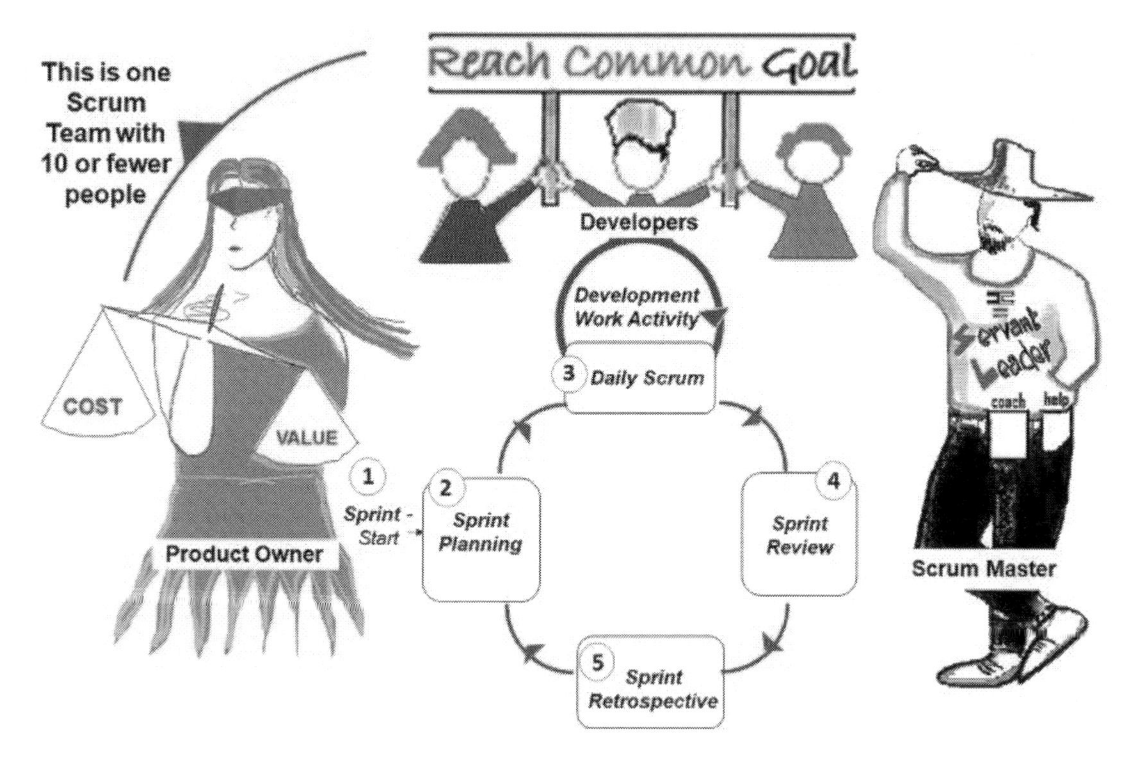

Fig. 7- Scrum Team

-------------------Question- 11-------------------------

The Scrum Master is the manager of the Scrum Team.

a) True

b) False

-------Answer-------

In Scrum, there is no exclusive team manager role like a Project Manager. There are project management activities in Scrum, but they are distributed among the three Scrum roles. The Scrum Master manages Scrum deployment and coaches the team on Scrum. Correct answer is 'b.'

-------------------Question- 11-------------------------

Is the Scrum Master only a teaching role?

Scrum Master is a management position in Scrum. The Scrum Master manages the Scrum Implementation. The Scrum Master coaches the Scrum Team to realize their potential. They are not just teachers or coaches. They are responsible for many other activities that are instrumental in transforming teams into value creators.

An example of a critical activity they play is when the Scrum Team runs into issues that prevent it from achieving their goal and if these issues are outside the team's influence, the Scrum Master owns these impediments and resolves them. The Scrum Master also shields the team from unnecessary external interruptions, by helping those outside the Scrum Team understand which of their interactions with the Scrum Team are helpful and which aren't. The Scrum Master sets the goals for the team to improve their way of working every Sprint.

The Scrum Master also helps the organization to adopt Scrum by helping everyone understand Scrum theory, practices, rules, and values.

Where is the manager for the project?

In a Scrum Team, there is no role other than the three previously mentioned roles. This means there is no Project Manager.

Building complex products is a knowledge game. The team involved in this game contains a knowledge-based workforce. If they face challenges in their work, they are knowledgeable enough to collaborate in innovative ways to address them.

However, this ground-level team needs to be structured and empowered to become self-managing and cross-functional.

One hindrance to self-management is how project teams are traditionally managed. Usually, a team is assembled with individuals having some unique skill needed for the product development. These individuals are expected to wait for tasks to be assigned to them by a manager. Managers command these individuals with direction and control. Such leadership is known as command and control leadership.

Once their task is completed, the individuals wait for the next task to be assigned. Instead of self-managing as a team to solve the challenges, they usually just limit their contribution at an individual level and execute what is asked.

Another pitfall with command and control leadership is that more often the project decision and directions will reflect the subjectivity of one commanding individual. But, in complex adaptive problems the changes and challenges are generally multi-dimensional. They require multi-dimensional analysis to comprehend and respond. For such multi-dimensional thinking and action, it is desirable to leverage the collective wisdom of all members of the team rather than just a commanding individual.

Therefore, there is no commanding and controlling manager within Scrum. The Scrum Team, which is the team on the ground that knows the reality, takes full ownership, and self-manages their work.

-------------------Question- 12-------------------------

In a traditional approach, <fill in> organizes and manages the team members' work, and <fill in> is their management style.

a) Business Manager, People Centricity

b) Project Manager, Command and Control

c) The team, Self-Management

-------Answer-------

Correct answer is 'b.'

-------------------Question- 12-------------------------

If there is no Project Manager, how do we engage existing Project Managers?

In a traditional model, the Project Manager controls the project budget, the project team members, and the project tasks. In Scrum, these activities are distributed between the three Scrum roles.

Though the Project Manager role is eliminated, there are still management positions within Scrum. The Developers collectively manage the project tasks and their own work. The Product Owner manages the business investment. The Scrum Master manages how Scrum is implemented.

As for the existing Project Manager, they can choose one of these management positions. However, they need to consciously choose the position with the understanding of the responsibilities involved. None of these management positions, such as Product Owner, Scrum Master, etc. involve managing people or unilaterally controlling the project plan or tasks. This will be further explained later in the book.

-------------------Question- 13--------------------------

In self-managing teams, the work is divided between individual team members. Each team member takes accountability for the progress of only their work.

a) True

b) False

-------Answer-------

In self-managing teams, the work is decomposed into work units. Individual team members take some work units. However, the entire team is accountable for the overall progress of reaching a common goal. Correct answer is 'b.'

-------------------Question- 13--------------------------

Self-Managing Teams balance their self-empowerment to create value

In a self-managing team, all the team members plan their work together and track its progress against a common goal. They also identify and resolve challenges together. Without any management or direction from outside, they strive to balance flexibility, creativity, and productivity, so they can maximize the collective value of their work.

-------------------Question- 14--------------------------

In their journey to deliver products of the highest business value, what factors will enable the Scrum Team to balance creativity, flexibility, and productivity? Select all that apply.

a) Strong Team Management and Guidance by a team member identified as their leader

b) Having all the skills required to perform all their work without external help

c) A Performance Management System that rewards the super achievers of the team

d) Structuring the team such that it can self-manage its work against a common goal

-------Answer-------

If sufficient capabilities and empowerment are not present, the team cannot acquire flexibility, nor it can command the creativity and productivity. Sufficient capabilities are ensured by having all the skills required for the job. Empowerment is ensured by the structure of self-management. Correct answers are 'b' and 'd.'

-------------------Question- 14--------------------------

Who manages the people aspects of the Scrum Team?

Every Scrum Team usually works within a larger organizational ecosystem. Though Scrum does not have management roles like Project Managers, there is a shared understanding among Scrum Practitioners about the important role "Managers of Organization" play. These managers set and manage larger strategies, define operational units, structure self-managing teams, and help to resolve organizational impediments to Agility.

For any personnel issues within the team's influence, such as personality conflicts, the self-managing team itself will resolve them.

If the issue is outside their influence, such as human resource management functions of hiring, firing, compensation, and other legal aspects, they are handled by the appropriate human resource authority defined by the organization's management.

-------------------Question- 15--------------------------

As a self-managing team, what can a Scrum Team manage? Select all that apply.

a) Managing their work to reach a common a goal

b) Managing their human-related aspects like leave, firing one of the team members, office dress code, etc.

c) Managing some other teams

d) Managing to support ad-hoc high-priority requirements from important executives

-------Answer-------

Self-managing teams are empowered to organize and manage their work. However, the team cannot self-manage the human resource related aspects. They also cannot work from a different set of requirements outside the Product Backlog. Correct answer is 'a.'

-------------------Question- 15--------------------------

How is progress measured in Scrum? What are the key performance indicators (KPI)?

Scrum is founded on empiricism where the decisions about the next Sprint are made based on the insight obtained from the previous Sprint.

Every Sprint must create at least one piece of functionality that is useable by the intended users. Only then can it lead to inspection, usage, and useful feedback. So, the Increment, which is a body of inspectable and usable outcome, is the only measure of progress in Scrum. There is no other metric of progress such as creation of any interim documents/artifacts, completion of phases, etc.

In addition, a Scrum Team may internally use some measures such as Sprint Work Planned vs. Completed (Burn-down), Rate of Completion (Velocity), etc. However, these are only internal metrics used by the team to manage their work. They are not indicators of progress for stakeholders.

-------------------Question- 16--------------------------

In Scrum, Team Velocity is a good metric to track the progress of product development.

a) Yes

b) No

-------Answer-------

In Scrum, the real mark of progress is the delivery of a useable product Increment (thereby bringing the product closer to the overall Product Goal) in every Sprint. Correct answer is 'b.'

-------------------Question- 16--------------------------

If there is no detailed project plan, how will we know the details of the project and work execution?

Scrum uses three artifacts to track the information about the Product and the work. We have already discussed two of the artifacts, the Increment, and the Product Backlog.

1. The <u>Increment, which is a body of inspectable and usable outcome,</u> is the mark of the real progress, and it provides information for required stakeholders about the progress so far.

2. The <u>Product Backlog</u> is the product definition in terms of ordered product features. It is never frozen (closed for changes) and is a continuously evolving artifact because the Scrum Team and Product Owner in particular are always looking for changes and opportunities to maximize the value of the product. Using the ordered features, one can understand what the team will work on in the future.

3. The <u>Sprint Backlog</u> is a temporary artifact created for each Sprint. It contains the Sprint Goal (why), subset of the Product Backlog Items chosen to be delivered in the upcoming Sprint (what), the plan of how to deliver them (how), and optionally one or more improvements required in the team's way of working. It is the plan maintained by the team about what tasks should be performed within the current Sprint.

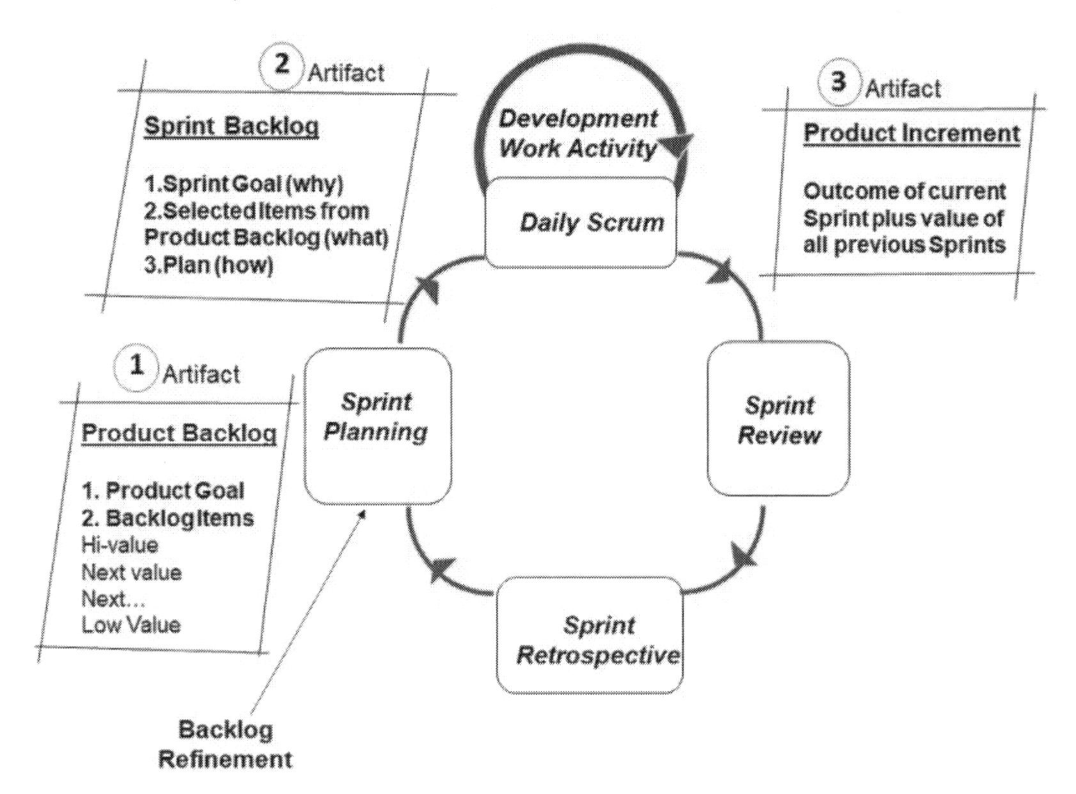

Fig. 8- Scrum artifacts

-------------------Question- 17-a-----------------------

A Scrum Team must produce the following artifacts. Select all that apply.

a) Project Plan
b) Product Backlog
c) Design Document
d) Sprint Backlog
e) Test Case Specifications
f) Project Status Report
g) Increment

-------Answer-------

A Scrum Team produces and maintains artifacts that help them to plan their work, track their progress, and share information visibly to required stakeholders. In that respect, Scrum mandates only three artifacts. Notice that the Increment is not a document but a body of inspectable and usable working product Increment. Correct answers are 'b,' 'd,' and 'g.'

Other artifacts or documents are optional and can be chosen by the team if they add value to their work or final product.

-------------------Question- 17-a-----------------------

Commitment to the Three Artifacts

Each of the three artifacts contain 'commitments' to them. For the Product Backlog it is the Product Goal, the Sprint Backlog has the Sprint Goal, and the Increment has the Definition of Done. They exist to bring transparency and focus toward the progress of each artifact.

-------------------Question- 17-b-----------------------

Scrum Teams create six artifacts: Product Backlog, Sprint Backlog, Increment, Product Goal, Sprint Goal, Definition of Done.

a) True

b) False

-------Answer-------

In Scrum, only the following three are called artifacts: Product Backlog, Sprint Backlog, and Increment. Other three (Product Goal, Sprint Goal, and Definition of Done) are reflections of the 'commitment' to these artifacts. Correct answer is 'b.'

-------------------Question- 17-b-----------------------

Summary

- Traditional command and control leadership blocks the team's bottom-up intelligence. Complex product building is a knowledge (not physical labor) driven work where the people closer to the work often have more knowledge and local information about the work than their bosses. So, it is important for the bosses (managers) to empower their ground-level team members to collectively decide (using the team intelligence) how to perform and troubleshoot the work without any external direction.

- Scrum facilitates team intelligence by building a small cross-functional team that self-manages its work.

- Cross-functional means having all competencies needed to accomplish the work without depending on others not on the team.

- Self-managing means empowerment to choose how best to accomplish their work (internally decide who does what, when, and how) rather than being directed by others outside the team.

- The Scrum Team consists of a Product Owner, the Developers, and a Scrum Master.

- The Developers build the Increment.

- Individual Developer may have specialized skills, but everyone's title is Developer.

- Individual Developer may have areas of focus, but the entire Scrum Team is accountable for creating a valuable and useful Increment every Sprint.

- The Product Owner is responsible for maximizing the value of the product and the work of the Scrum Team.

- The Scrum Master is responsible for promoting and supporting Scrum as defined in the Scrum Guide. Scrum Masters do this by helping everyone understand Scrum theory, practices, rules, and values.

- The Scrum Master is a servant-leader for the Scrum Team. The Scrum Master helps those outside the Scrum Team understand which of their interactions with the Scrum Team are helpful and which aren't. The Scrum Master helps everyone change these interactions to maximize the value created by the Scrum Team.

- The team model in Scrum is designed to optimize flexibility, creativity, and productivity.

- The Scrum Team uses three artifacts that are specifically designed to maximize transparency of key information about the product and the work so that everybody has the same understanding.

- Each artifact contains a commitment to ensure it provides information that enhances transparency and focus against which progress can be measured. For the Product Backlog it is the Product Goal. For the Sprint Backlog it is the Sprint Goal. For the Increment it is the Definition of Done.

#

Chapter 3 – Scrum - A container and collaboration framework

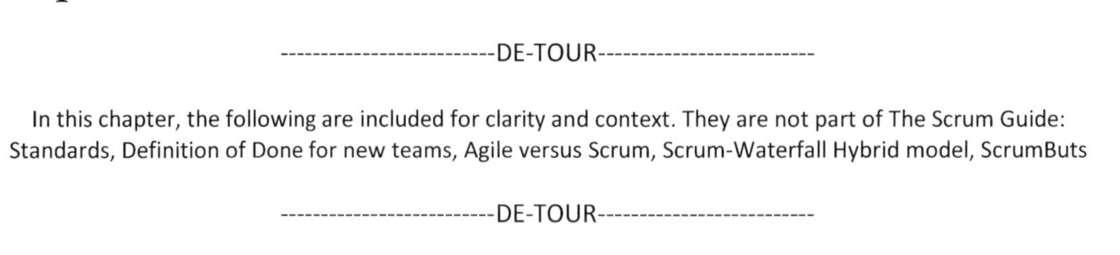

-------------------------DE-TOUR-------------------------

In this chapter, the following are included for clarity and context. They are not part of The Scrum Guide: Standards, Definition of Done for new teams, Agile versus Scrum, Scrum-Waterfall Hybrid model, ScrumButs

-------------------------DE-TOUR-------------------------

Scrum is not a methodology, process, or technique for building products. It is a framework within which one can employ various product building processes and techniques. Scrum is a collaboration framework within which people can address complex adaptive problems, while productively and creatively delivering products of the highest possible value.

As a framework, Scrum provides a broad structure consisting of a Scrum Team and associated roles, events, artifacts, and rules. Scrum is lightweight with only three roles, five events, and three artifacts. It is simple to understand. This structure enables a simple but effective way of working together as a team towards a focused goal.

Scrum is NOT a process but a framework – What does this mean?

Scrum is not a process or methodology for building products. Unlike a process or method, it does not prescribe a detailed development blueprint specific to an industry sector or domain.

It is a container framework wrapping around any appropriate process or technique. A team within an industry sector can choose to employ industry specific processes and techniques within Scrum. For example, the software building Scrum Team can employ software engineering techniques such as continuous integration, Test Driven Development, etc. as part of their development work.

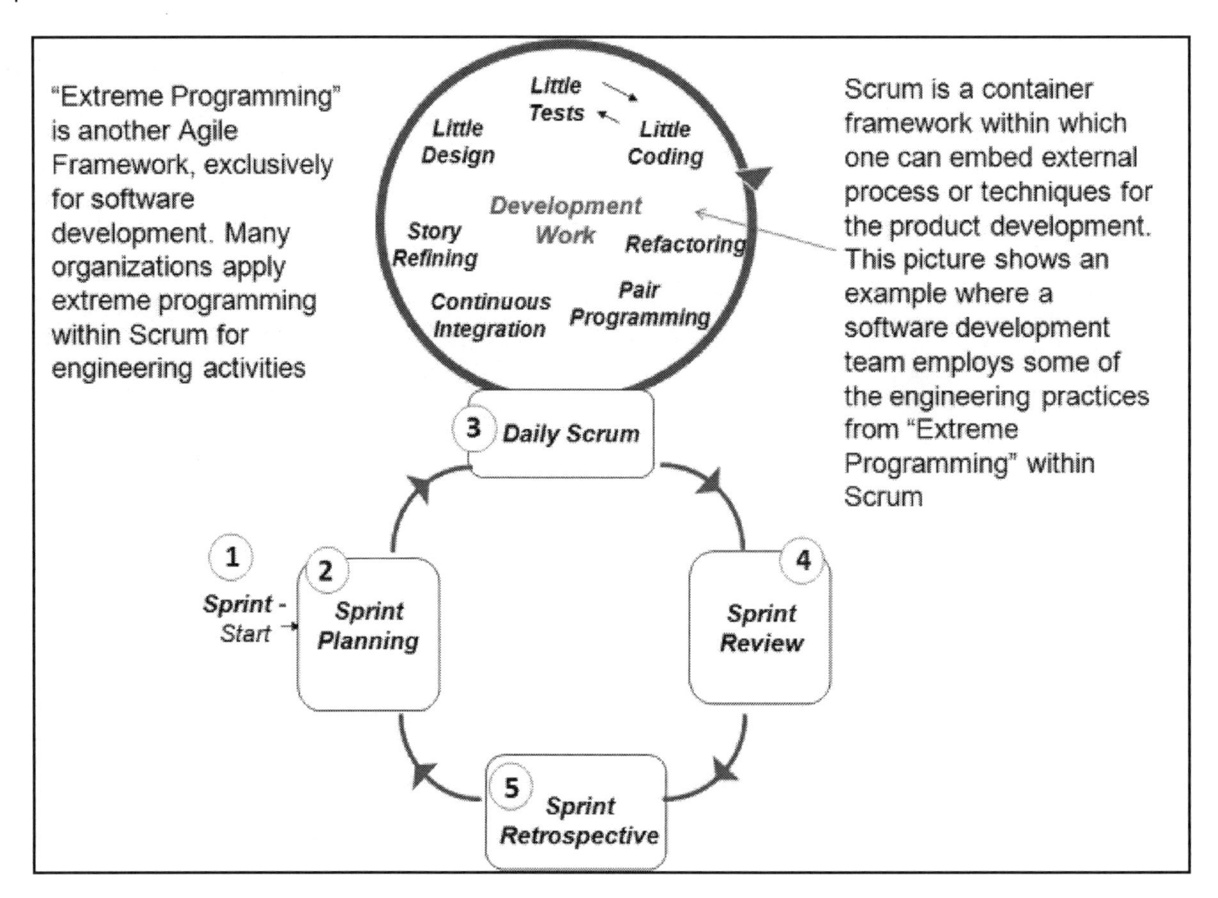

Fig. 9- Scrum- Container of other Processes and Techniques

-------------------Question- 18-------------------------

Scrum is best described as a

a) Software methodology.
b) Framework for developing and sustaining complex products.
c) Product development process.
d) Collection of best practices.

-------Answer-------

Scrum is a framework within which appropriate processes and techniques can be employed to develop complex products. Correct answer is 'b.'

-------------------Question- 18-------------------------

How does this framework work? - Heartbeat of Scrum

Irrespective of the domain-specific product building techniques applied by different Scrum Teams, all teams follow the same Scrum framework. Scrum is founded on empiricism, and three pillars uphold every implementation of empiricism. Scrum events are built around these three pillars. If these pillars are properly followed, Scrum will be healthy.

The three pillars of empiricism are

1. Transparency: Transparency means

- Providing visibility of information about the work and the outcome.
- Using common standards for information so that observers will share the common interpretation and understanding.

By transparency, the significant aspects of the work must be visible to those responsible for the outcome.

2. Inspection: performing frequent reviews of the Scrum artifacts and progress towards the Sprint Goal to get early feedback on undesirable variances.

3. Adaptation: performing adjustments if the inspection finds variance beyond acceptable limits, and hence the resulting product will be unacceptable. The Scrum Team has collective responsibility to make the adjustments as soon as possible to minimize further deviation.

-------------------Question- 19-------------------------

Transparency, Inspection, and Adaptation are the three pillars of

a) Empirical Process Control Theory.
b) Lean.
c) PDCA.
d) Six Sigma.

-------Answer-------

Correct answer is 'a.'

-------------------Question- 19-------------------------

Other than the Sprint itself, which is a container for all other events, each of the other four events in Scrum is a formal opportunity to inspect and adapt. These four events are predefined points of inspection to understand what has happened. In every Scrum event where inspection is performed, there can be an opportunity to adapt and respond. Fig. 10 shows the events of Transparency, Inspection, and Adaptation.

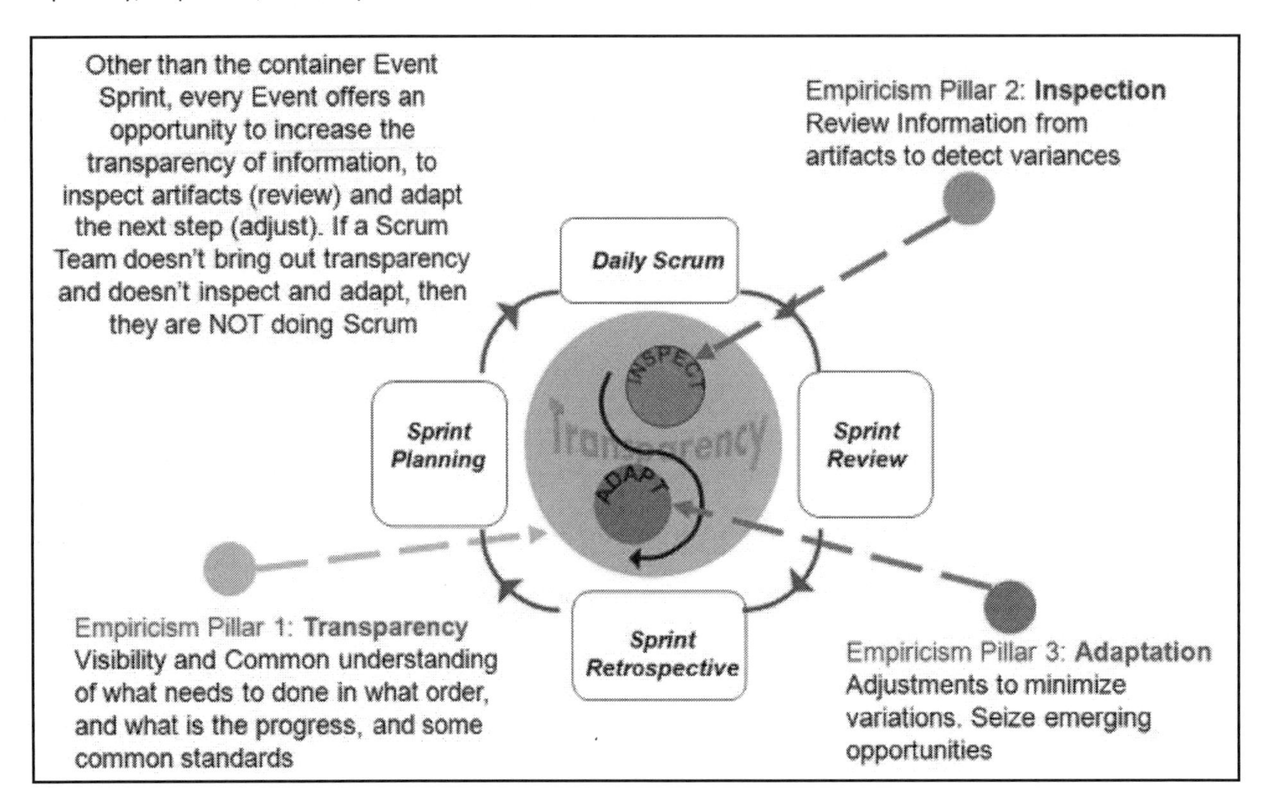

Fig. 10- Pillars of Empiricism

What is an example of the team applying a transparency principle?

The empirical approach requires the team to increase the transparency of the information as much as possible. One can increase the transparency by keeping the information factual, making it visible to those responsible for the outcome, and establishing common standards. An example of a common standard is the Definition of Done. Those performing the work and those accepting the work product must share a common Definition of Done.

------------------Question- 20-------------------------

Transparency in empiricism refers to

a) Clear thinking and planning by each team member.

b) The significant aspects of the product development process are defined by common standards and made visible so the observers will share the same understanding.

c) The highest levels of morality.

-------Answer-------

Transparency requires that significant aspects of the process must be defined by common standards, and they must be visible to those responsible for the outcome. Correct answer is 'b.'

------------------Question- 20-------------------------

--------------------------DE-TOUR--------------------------

'Standard' is not a formal element of the Scrum framework. To add clarity, this book uses the term 'Standard' to denote things like Definition of Done, Definition of Ready, etc. Starting with 'The Scrum Guide 2020,' Definition of Done is called 'commitment' to the Increment. This book will use both 'standard' and 'commitment' interchangeably to refer to Definition of Done.

--------------------------DE-TOUR--------------------------

The Definition of Done defines set of conditions that must be met in order to accept the team's Sprint outcome as a product Increment. In other words, a product Increment is a body of inspectable, Done work.

As an example, the following can be some possible conditions of a Definition of Done:

- The Completed Product Backlog Item must pass all automated integration tests.
- The Completed Product Backlog Item must have an associated technical document listing impacted technical components.
- The Completed Product Backlog Item must meet the technical performance requirements defined in the organization's performance objectives.

By using the Definition of Done, everyone transparently understands what a Done Product Backlog Item or a Done Increment means. In the Sprint Review, a Product Owner will accept a Product Backlog Item as complete if and only if it meets the conditions set forth in the Definition of Done. The Definition of Done mostly contains technical conditions such as quality, performance, etc.

-------------------Question- 21--------------------------

What is used by the Scrum Team to identify unfinished work in a Sprint?

a) Coding Standard
b) Definition of Ready
c) Testing Standard
d) Definition of Done

-------Answer-------

The Definition of Done provides the same shared understanding of what it means for the work to be complete. It spells out all that is required to get the work 'done.' So, it is used by the team to assess what is yet to be done to complete the product Increment. Everybody developing the same shared understanding is the key to transparency. Correct answer is 'd.'

-------------------Question- 21--------------------------

The Definition of Done need not be the same between different Scrum Teams of an organization. However, any one product or system should have one Definition of Done which will be a standard for any work done on it. This will be discussed further in Part 2.

The Scrum Team lives by Scrum Values so that the Scrum Pillars come to life

Scrum values are a set of fundamental qualities underpinning the Scrum framework. The older versions of The Scrum Guide did not contain these values. Later the authors regarded these values as an important common denominator to develop better software and hence added them to The Scrum Guide. Scrum Teams live by five values: commitment, courage, focus, openness, and respect. Being proficient in living these values brings the Scrum pillars of Transparency, Inspection, and Adaptation to life and builds trust for everyone. The Scrum Team members learn and explore these values as they work with the Scrum events, roles, and artifacts. These values are useful as another checkpoint to compare the behavior within the Scrum Team, i.e., if the behavior reflects the understanding of the values or just the mechanics (following only rituals).

The five values of Scrum are:

1. <u>Commitment</u>

Commitment of every team member to achieve the goals of the Scrum Team. Commitment in following the pillars of empiricism, lean thinking and self-management and using them to achieve the goals.

2. <u>Courage</u>

Courage to work on tough problems. Courage to do the right thing by accepting that the future cannot be predicted and responding to emerging change is the way forward. Courage helps everyone to be grounded (in reality), and not giving into personal pride.

3. <u>Focus</u>

Focus of the team on prioritizing and completing the Sprint work to achieve the goals of the Scrum Team. Focus helps to avoid doing other things not related to the Sprint Goal.

4. <u>Openness</u>

Openness of the Scrum Team and its stakeholders in expressing and facing the facts and truths about all the work and challenges with performing the work, thereby increasing transparency. Openness to collaborate with others with the highest amount of transparency.

5. <u>Respect</u>

Respect each other as capable and independent people so that it can provide a trustworthy environment to learn and share.

-------------------Question- 22-------------------------

A Developer is requested by an important stakeholder to help them with an urgent and important task outside the Sprint Goal. The Team member set aside the Sprint work for the day and instead helped with this request. Which statement best describes the Team member's action?

a) The Team member has gone the extra mile and must be rewarded.

b) The Team member has violated Scrum rules by not consulting with his manager.

c) The Team member did not live by the Scrum value of focus.

-------Answer-------

The Scrum value of focus helps to avoid doing other things not related to the Sprint Goal. The Team member is expected to live the Scrum value of focus by prioritizing and completing the Sprint work to achieve the goals of the Scrum Team. Correct answer is 'c.'

-------------------Question- 22-------------------------

What if our team is not mature enough to create a useable Product Increment every Sprint?

The objective of every Sprint is to produce at least one valuable and useful Product Increment. The Definition of Done should have conditions (quality measures) that the Product Increment must meet. For newer teams this is often a big challenge.

Yet the Definition of Done should not be set with diluted quality measures with the objective of making it easy to meet. Unless the Increment is of useful quality, the Scrum Team cannot get feedback from actual usage. Diluting the Definition of Done will hide the current weaknesses in Product Development.

Given this, even a new team should define Done with conditions such that the Increment will be valuable and useful. At the same time, the conditions need to be realistic to motivate the team. Over the iterations, as the team's ability matures, more stringent conditions can be gradually added. Having a realistic Definition of Done for a new team means that the working Increment may have known bugs, but they are transparent between the Developers and the Product Owner.

--------------------------DE-TOUR--------------------------

Since Scrum was in existence before the Agile movement, Agile is not referred to within Scrum. Today Scrum is widely seen as one of the "methods" under the broader umbrella of "Agile." Many regard Scrum and Agile as being the same. We need to put these things into perspective before moving on.

--------------------------DE-TOUR--------------------------

How does Scrum relate to Agile?

Agile within software development is associated with "The Agile Manifesto." The Agile Manifesto is a proclamation of a better way of working to create software. The Agile Manifesto is a set of values and principles for a new way of software development. Scrum has contributed a lot to the development of Agile. See agilemanifesto.org for more information.

Though the Agile Manifesto is widely seen as the mother of all Agile-based frameworks, Scrum, which is an alternative software development model, existed before the Agile Manifesto was written.

Scrum started as an alternative approach to complex product development several decades back. The rough idea of Scrum in product development was introduced by Hirotaka Takeuchi and Ikujiro Nonaka in their white paper titled "The New New Product Development Game," which was published in the Harvard Business Review in 1986.

Ken Schwaber and Jeff Sutherland introduced Scrum as an alternative to traditional development models to systems and the software development world. They presented a process framework called Scrum at the 1995 OOPSLA Conference. They presented Scrum as an enhancement to traditional models of systems development.

Later they defined the Scrum framework that employs Scrum Teams and the associated roles, events, artifacts, and rules to produce frequent working Product Increments.

Scrum is a standalone framework, but it respects the Agile Manifesto

The Agile Manifesto was written by group of representatives of "alternative implementations of software delivery models" in February 2001. The authors of The Scrum Guide (Ken Schwaber and Jeff Sutherland) were among these representatives.

In principle, the Agile Manifesto's ideas have a lot in common with the Scrum framework elements. Scrum mutually respects the Agile Manifesto values and principles. Scrum explicitly lists "Understanding and practicing agility" as one of the services that the Scrum Master provides to a team.

Agile is a philosophy about a "Newer way of developing software." It is a philosophy because it is not prescriptive on an exact implementation. Scrum is one of those concrete implementation frameworks to help people develop any complex product not just software. The Scrum framework definition is concrete with Scrum Teams and the associated roles, events, artifacts, and rules.

We want to become Agile. Where do we start?

Anyone wanting to transition to Agile should understand the Agile Manifesto, and its values and principles. Many organizations embrace the Agile values and principles at the conceptual level. Then they decide on a concrete implementation framework such as Scrum that gives a structure to the Agile way of working. After that, the Scrum Team employs additional techniques that add value specifically to them within the Scrum framework. For example, many Scrum Teams in the software domain employ Extreme Programming (XP) practices within the Scrum framework to add agility to their development work.

Original Scrum is defined in The Scrum Guide - Immutable

For organizations with historical development practices and infrastructure, the most common scenario after applying Scrum is that the Scrum Teams may encounter issues that will impede their effort to create valuable and useful Increments within short Sprints. Scrum will expose such dysfunctions in the current organizational ecosystem. It is normal and expected.

The organizations should try to correct these dysfunctions. Sometimes organizations take the route of ScrumButs to handle these dysfunctions.

ScrumBut refers to an adjustment or modification made to Scrum so that the organization can hide the problem instead of addressing it.

Scrum.org defines ScrumButs as having a particular syntax:

(ScrumBut)(Reason)(Workaround)

Scrum.org provides some examples of ScrumButs:

"(We use Scrum, but) (we cannot build a piece of functionality in a month,) (so our Sprints are 6 weeks long.)"

"(We use Scrum, but) (sometimes our managers give us special tasks,) (so we do not always have time to meet our Definition of Done.)"

Hiding the weaknesses using ScrumBut will take away the opportunity for organizations to address them and become agile.

-------------------Question- 23-------------------------

Shortly into using Scrum for the first time in an organization, the Scrum Team runs into several issues against following Scrum. The most common inference is

a) Scrum should have been applied for Product Development instead of Software Development.

b) The team should have followed only the Scrum's guidance about how to perform software engineering practices like design, coding, testing, etc.

c) The Scrum Team didn't plan the product development project completely in advance.

d) It is normal for first timers. Scrum will expose all weakness in the current organizational ecosystem. They should be treated as the opportunities for improvement and need to be resolved.

-------Answer-------

For organizations with historical development practices and infrastructure, the most common scenario after applying Scrum is that it will expose the weaknesses in the current organizational ecosystem. It is normal and expected. The organizations should strive to address these weaknesses while maturing their team's ability to produce useable software within the Sprints. Correct answer is 'd.'

-------------------Question- 23-------------------------

Can we follow a hybrid approach of combining Scrum and Waterfall?

Some organizations learn about the dramatic changes that Scrum brings and want to implement Scrum. At the same time, they want to implement Scrum "smoothly." A common approach is to follow a hybrid model where they retain the existing methods and selectively apply partial Scrum. An example is the planning of Sprints into a Design Sprint, Development Sprint, Testing Sprint, etc. This is nothing but Waterfall in a Scrum disguise.

The Scrum authors are particular about providing a version of Scrum that will maximize the benefits intended. The approved body of knowledge on Scrum is "The Scrum Guide." The authors position this version of Scrum in The Scrum Guide as immutable, i.e. it cannot be adulterated with customized practices or hybrid terminology.

Any such act will dilute the identity and distinguishing character of Scrum. The adulterated Scrum without its original identity may not be seen as a "change for the better" by the people in the organization.

If someone claims that a hybrid model is working fine, they need to investigate if they can witness the transformational changes that Scrum brings:

- *Has the risk of creating waste gone down by early and frequent delivery of production-quality Increments?*
- *How many investments were identified as "waste or questionable" and severely modified or terminated early?*
- *Are the Scrum Teams producing more valuable and successful products?*
- *Has the engagement with business partners and customers increased significantly?*

------------------Question- 24-------------------------

Is Scrum immutable?

a) Yes

b) No

-------Answer-------

Changing Scrum or customizing it for the convenience of an existing culture may dilute its distinguishing identity as a "change agent." Also, it may be perceived as just another additional practice which fails to motivate those who anticipate change. Correct answer is 'a.'

------------------Question- 24-------------------------

We already have multiple releases. Why should we consider Scrum?

Some organizations may already have multiple releases of their product instead of batching. However, they may not have a disciplined product development approach to contain the risks and increase the value.

- Scrum enforces the discipline of timeboxing: A Scrum Team produces a valuable and useful Increment every few weeks. Timeboxing into a few weeks leads to frequent, highly relevant feedback that enables the teams to replan and mitigate risks. At the end of a few weeks, if the review reveals that the Increment is a waste, the direction for the next Sprint can be adjusted. The risk of pursuing a wrong direction is limited to the cost of one Sprint.

- Scrum nurtures the owner's mindset at the team level: Scrum offers several opportunities to course correct and make the right decisions. To make appropriate decisions, the team must have an owner's mindset. In Scrum the Product Owner is given empowerment on product decisions, and the team is given empowerment on their work decisions.

Some organizations have occasions where such self-managing teams release the initial Increments to actual use (without waiting for the full product suite) and realize business benefits early. Since the product is still being built, these early benefits then can be used to fund the subsequent work.

In some other instances, by completing the high-value items in the Product Backlog, the team may have realized most of the intended business value already. In such cases, the product building effort can be closed early, thus saving cost. This disciplined ecosystem of risk mitigation through empiricism, bottom-up intelligence, and the owner's mindset of value maximization is possible only with the application of Scrum in its entirety.

Summary

- Scrum is NOT a process for building products.
- It is a container framework within which one can employ various processes and techniques.
- It is a collaboration framework within which people can address complex adaptive problems while productively and creatively delivering products of the highest possible value.
- It focuses on value delivery and may not reflect a traditional project approach.
- It has only three accountabilities – Product Owner, Scrum Master, Developers.
- It has only three artifacts – Product Backlog, Sprint Backlog, Increment. There is a commitment attached to each of the artifact: Product Goal, Sprint Goal, Definition of Done.
- It has only five events – Sprint, Sprint Planning, Daily Scrum, Sprint Review, Sprint Retrospective.
- The three pillars of empiricism are the heartbeat that upholds every implementation of the empirical process control.
- The pillars are Transparency, Inspection, and Adaptation.
- Transparency requires that significant aspects of the process be visible and defined by a common standard.
- The Definition of Done is a standard for ensuring Transparency. Its definition must enable the team members to have a shared understanding of what it means for the work to be complete.
- Inspection requires that Scrum users frequently inspect the Scrum artifacts and progress towards a Sprint Goal to detect undesirable variances.
- Adaptation requires that, in the event of unacceptable variances, the Scrum Team must make adjustments as soon as possible.
- The Scrum Team members learn and explore five values of Scrum as they work with the Scrum events, roles, and artifacts. When the values of commitment, courage, focus, openness, and respect are embodied and lived by the Scrum Team, the Scrum pillars of Transparency, Inspection, and Adaptation come to life and build trust for everyone.
- Scrum existed before Agile. Scrum respects the Agile Manifesto.
- Implementing only parts of Scrum is not Scrum. Scrum is immutable.

#

Quiz 1

1. The organization or senior management's support

a) Is not needed for Scrum implementations.

b) Is not needed because there is no scope for management in Scrum.

c) Is needed to support the Product Owner to maximize the product value and the Scrum Master to coach and implement Scrum.

-------Answer-------

The organization or senior management's support is a critical success factor for product planning and development in an empirical environment. Their action is needed to structure the self-managing teams. Correct answer is 'c.'

2. A Scrum Team is at the end of a Sprint. The next Sprint starts

a) Only after the product Increment is released to production.

b) Only after the Retrospective event of the current Sprint.

c) Only after the team for the next Sprint is on board.

d) Only after the Sprint Planning.

-------Answer-------

The Product Owner may choose to release the Increment to production, but it is not mandatory. The same team will continue to work on the next Sprint. Sprint Planning is the first event of the Sprint. The last event of the Sprint is the Retrospective. Correct answer is 'b.'

3. Only the Product Owner can come up with items that can be considered for the Product Backlog. Others cannot provide input/recommendations/ideas about new items.

a) True

b) False

-------Answer-------

While the Product Owner has the final say on the content and order of the Product Backlog, they can get the input/ ideas about new items from any stakeholder or Scrum Team member for consideration. Correct answer is 'b.'

4. A Scrum Team needs to have the following roles. Select all that apply.

a) Product Manager

b) Business Manager

c) Architect

d) Developer

e) Tester

f) Programmer

g) Scrum Master

h) Project Leader

i) Product Owner

-------Answer-------

Scrum contains only three accountabilities. Anyone working in development, irrespective of their specialized skills such as architecture, testing, UI designing, coding, technical documenting, etc. is called a Developer. Correct answers are 'd,' 'g,' and 'i.'

5. A Scrum Team must produce the following artifacts. Select all that apply.

a) Project Plan

b) Product Backlog

c) Design Document

d) Sprint Backlog

e) Test Case Specifications

f) Project Status Report

g) Increment

-------Answer-------

A Scrum Team produces and maintains artifacts that help them to plan their work, track their progress, and share the information visibly to required stakeholders. In that respect, Scrum mandates only three artifacts. Notice that the Increment is not a document but a body of inspectable and usable working product Increment. Correct answers are 'b,' 'd,' and 'g.'

Other artifacts or documents are optional and can be chosen by the team if they add value to their work or the final product.

#

Part 2- Scrum descriptive for Professionals

Over the long term - we overestimate what we can do. Over the short term - we underestimate all that is possible. To achieve more in the short term, use the magic of timeboxing with preset deadlines. The deadlines will push you into action from paralysis.

#

Chapter 1 – How to start Scrum

--------------------------DE-TOUR--------------------------

In this chapter, the following are included for clarity and context. They are not part of The Scrum Guide:
Staffing plan, Project Management aspects, Upfront architecture, Definition of Ready, Spike

--------------------------DE-TOUR--------------------------

Scrum is a container framework with a focus on collaboration within which people can address complex adaptive problems while productively and creatively delivering products of the highest possible value. Building a product using Scrum is a discovery process that starts with just enough preparation. This may sound odd to those who have planned "Projects in traditional methods." In a traditional approach, a lot of preparation is required before the team starts the development work.

Traditionally, the Project Manager needs to forecast the budget, schedule, staffing plan, risk management plan, quality plan, and communication plan based on the project scope.

Scrum does not take such an approach. It does not attach value to plans and artifacts that are based on long-term assumptions. It starts with just enough preparation and pursues a "value discovery and maximization" journey. Therefore, Scrum is a journey and cannot be called a "Project" in the traditional sense.

What is the input to the first Sprint?

The inputs to the first Sprint are the Product Backlog and projected Sprint capacity of the Developers. The Product Backlog may contain only the initial business ideas. The projected capacity of the team is only a guess at this point. It will be refined over the upcoming Sprints as clarity emerges based on past performance.

Fig. 11 shows the lifecycle of Scrum. The journey starts with "just enough preparation." The journey concludes when enough value is delivered, the investment becomes unjustified, or when allocated resources are exhausted.

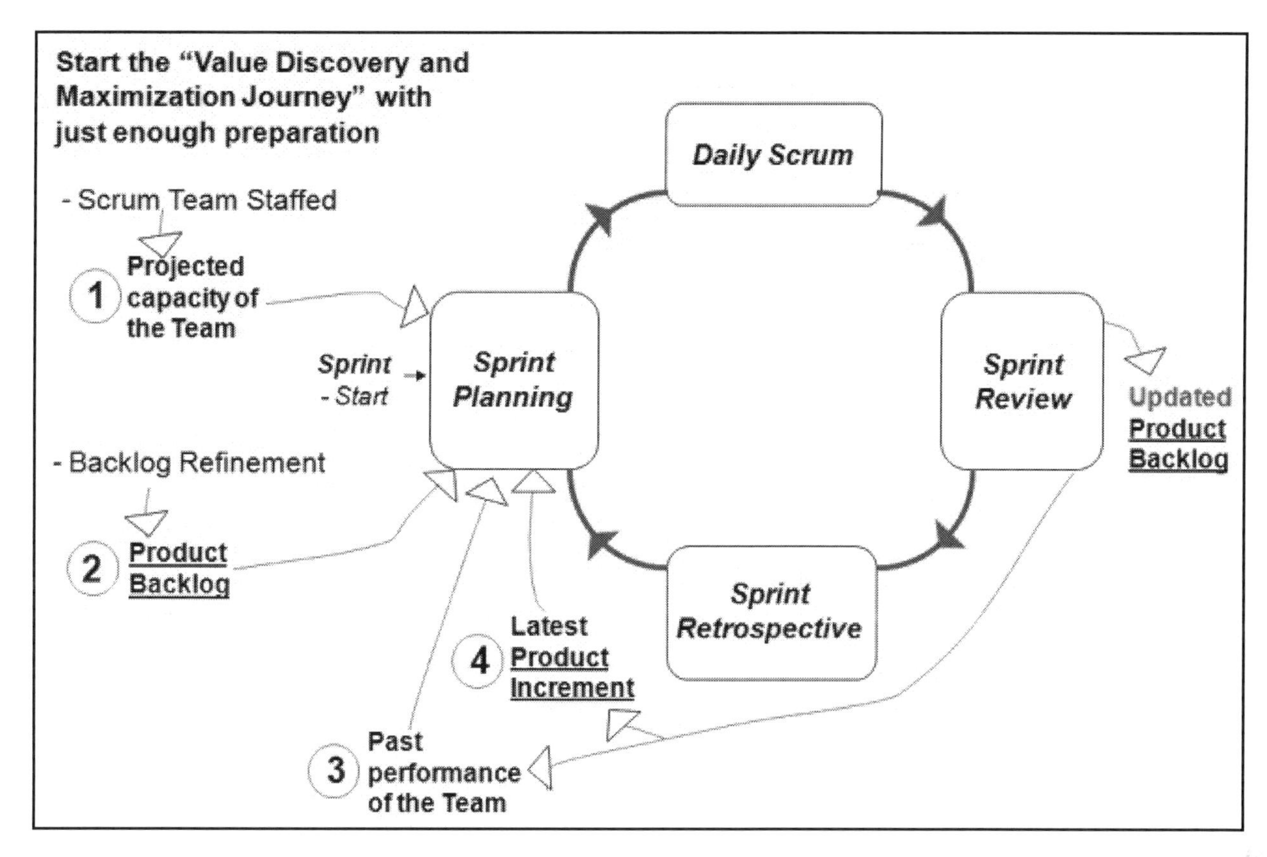

Fig. 11- Starting Scrum - Inputs to the Sprint

-------------------Question- 1-a------------------------

Select all that apply. Before starting the first Sprint, what needs to be in place?

a) A complete Product Backlog capturing detailed product needs.

b) Availability of the Project Manager.

c) Just enough Product Backlog Items with business ideas for the first Sprint.

d) Completed System Architecture.

e) Staffed Scrum Team.

-------Answer-------

There are no preconditions for the first Sprint. The availability of a Scrum Team and a list of business ideas for the first Sprint are enough to start the Sprint. Correct answers are 'c' and 'e.'

-------------------Question- 1-a------------------------

What about the input to subsequent Sprints?

Apart from the first Sprint, every other Sprint has more input to Sprint Planning. In addition to the Product Backlog and projected Sprint capacity of the Developers, the additional inputs to Sprint Planning include the latest product Increment and the past performance of the Developers.

How to staff without a staffing plan?

Traditional Projects bring people to the work and manage staffing complexities: In 'traditional projects,' the Project Manager produces a staffing plan based on the predicted work. This staffing plan projects the volume of work on a timescale, e.g. weekly or monthly. Using this projection, the "human resources" are brought to the work and removed when the work is complete.

The Project Manager manages and controls the complexities associated with maintaining this projection, forecasting the resource expenditure, and monitoring the full utilization of the resources.

<u>Scrum brings the work to a constant team and avoids staffing complexities</u>: The Scrum Team size is typically 10 or fewer people. It is small enough to remain nimble and large enough to complete significant work within a Sprint, with Developers, a Product Owner, and a Scrum Master. Before the Sprint starts, this team is staffed with all three accountabilities. It is staffed such that it has all the skills needed to create the Product Increment.

-------------------Question- 1-b------------------------

The size of the Scrum Team is

a) 7 plus/minus 2.
b) typically 10 or fewer people.
c) 3 – 9.
d) none of the above.

-------Answer-------

In the previous versions of The Scrum Guide, size of the Scrum Team was defined as 3 – 9 plus a Scrum Master and a Product Owner. From 2020 version, the language around the size is made more flexible- 'The Scrum Team is typically 10 or fewer people.' Correct answer is 'b.'

-------------------Question- 1-b------------------------

Once a Scrum Team is staffed, the team is maintained as a constant talent pool that is available throughout the journey. After establishing the constant team, Scrum facilitates 'bringing the work to this team.' The Product Owner brings the work, and the Developers map their availability to do the best-valued work. This self-management of the team eliminates the necessity for any manager to map their work.

There could be an occasional churn of team members going out and replacements coming in. There may be occasions where the team identifies a lack of some skills and will induct more developers with those skills.

So, other than the initial staffing and ongoing adjustments to maintain it, there is no need for an exclusive staffing plan.

-------------------Question- 2--------------------------

When a Scrum Team adds new team members to replace outgoing members, the productivity of the team

a) Will be negatively impacted.
b) Will be positively impacted.
c) Will remain the same.

-------Answer-------

When new team members join, the productivity of the team will be **temporarily** reduced. Correct answer is 'a.'

-------------------Question- 2--------------------------

What about other management aspects such as Scope, Quality, Risk, Communication, etc.?

All these aspects are owned, managed, and controlled within Scrum. However, there is no exclusive project management plan or person who manages it. These project management activities are distributed among the three Scrum roles.

Product Scope, Budget, and Schedule: The Product Owner manages the product scope in terms of the Product Backlog. They also manage the cost benefit of the product features. The cost of the work does not fluctuate as Scrum has a constant team. So, it is the Product Owner's responsibility to put together the features such that maximum value is obtained from this constantly available team. The Product Owner keeps the estimated time updated on the incomplete features. Keeping this up to date is important to forecast the schedule and completion timelines.

Quality Plan: The Scrum Team as a whole owns the quality of the Increment. The expected functional quality is specified in the tests. The expected technical quality is made transparent by the Definition of Done. The Scrum Team ensures the following rules:

- During a Sprint, the quality goals do not decrease.

- During each Sprint Retrospective, the Scrum Team plans ways to increase product quality by adapting the Definition of Done as appropriate.

- As a Scrum Team matures, it is expected that their Definition of Done will expand to include more stringent criteria for higher quality.

Risk Plan: The Scrum framework is fundamentally a risk reduction framework. It reduces the risk of big commitments, accumulation of waste, hidden weaknesses in product development abilities, etc. by limiting the planning horizon to shorter periods of time. Sprints enable predictability by ensuring inspection and adaptation of progress towards a Sprint Goal at least every calendar month. Sprints also limit risk to one calendar month of cost. Any other ongoing risks such as potential undone work are publicly communicated via impediments and acted upon by the team.

Communication Plan: Scrum defines clear boundaries for communication to increase the transparency and focus on valuable work.

The Product Owner is the single point of communication for stakeholders including customers, users, and management on product-related items. The Scrum Master manages the communication at the periphery of the team, such as multiple departments of the organization vs. the Scrum Team. The Scrum Master helps those outside the Scrum Team understand which of their interactions with the Scrum Team are helpful and which aren't. The Scrum Master helps everyone change these interactions to maximize the value created by the Scrum Team.

Within the Scrum Team, the communication is continuous. It is achieved by the transparency of information through artifacts. These artifacts provide necessary information about the product plan, work pipeline, current completion status, etc. Scrum events enhance the communication. In particular, Daily Scrums improve communication, eliminate other meetings, identify impediments to development, highlight and promote quick decision-making, and improve the Scrum Team's level of knowledge. This is a key inspect and adapt meeting.

Standards such as Definition of Done, Definition of Ready, coding standards, etc. establish the same understanding to all stakeholders about the expected work.

How can development start without an architecture or design?

Traditional projects create an upfront design: Traditionally, the entire technical design is created before development begins. This design is based on the current understanding of the business needs and technical solution patterns used in the past.

Scrum Teams emerge the design throughout the journey: Scrum Teams do not create a big upfront design before they start Sprints. Instead they evolve the design.

- You may recollect that the Product Backlog Items are refined in the Product Backlog Refinement sessions ahead of Sprints. The items are refined until they are transparent enough to estimate and confirm that they can be Done within a Sprint. To find the work estimate, the Scrum Team needs to visualize at least a simple skeleton design.

- In the Sprint Planning, this skeleton design is revisited and evolved to a finer level.

- Further, the Scrum Team usually chooses to set aside core design hours during the Sprint. As they get more clarity during the work, they evolve the best designs through continuous refactoring of initial designs.

- When some design patterns appear repeatedly throughout the Sprints, those patterns are 'locked in.' Most likely, such patterns may become the pillars of that system architecture.

As stated earlier, Scrum is a container framework within which Scrum Teams can employ domain-specific engineering techniques. The guideline above is one approach under the concept of "Design Emergence." Many teams also follow other variations. The skills required to emerge the design through constant refactoring is a separate subject.

-------------------Question- 3--------------------------

Select all that apply. In Scrum, the technical design of the solution is

a) Built one module after another with the Architect's guidance.

b) Initially created as a common architectural pattern by selected designers and architects and shared with others to build on top of it.

c) Started with just enough design which emerges throughout the Sprints.

d) Reached through focused attention during core design hours in the Sprint.

-------Answer-------

There is no Designer or Architect role in Scrum. Correct answers are 'c' and 'd.'

-------------------Question- 3--------------------------

How can the Scrum Team prepare enough Product Backlog Items before the first Sprint?

While there are a couple of ways to come up with an initial Product Backlog, it is not mandatory to have any minimum criteria for the Product Backlog before the first Sprint starts. All that is required for the first Sprint to start is a staffed Scrum Team and a set of business ideas to deliver in the first Sprint.

Approach I - A Scrum Team can initially work outside the Scrum Sprints to create and refine just enough of a Product Backlog to start. It should be made transparent to stakeholders that this activity does not lead to any opportunity of inspecting a useful outcome. Since this activity does not create working functionality, this activity should not be incorrectly called a Sprint. The time taken to arrive at this type of Product Backlog should be as minimal as possible.

Approach II - Start the Sprint Planning and refine just enough Product Backlog Items for the first Sprint. The team can craft the Sprint Goal, and then come up with a work plan for the initial days of the Sprint.

Once the Sprint is in motion, the Product Backlog Items are further refined in the Product Backlog Refinement sessions during the Sprint, so that there will always be some refined items ahead of future Sprints.

-------------------Question- 4--------------------------

A Scrum Team can have an exclusive first Sprint to prepare a Product Backlog, which is the sole outcome from that Sprint.

a) True

b) False

-------Answer-------

A Scrum Team can initially work **outside** the Scrum Sprints to create and refine just enough of a Product Backlog so that the first Sprint can start. However, this initial effort should not be called a Sprint. Also, it should only take a few days. Correct answer is 'b.'

------------------Question- 4-------------------------

There is no Sprint Zero or Iteration Zero before the first Sprint

Every Sprint must produce valuable and useful business functionality. Some teams create Sprint Zero for preparing a Product Backlog and other upfront preparation including tasks such as setting up the work environment, configuring tools, etc.

Sprint Zero does not produce any useable functionality. If there is no useable functionality, there is no opportunity for feedback on inspection and adaptation. Without the feedback, we will not know if the investment in the concluded Sprint was justified or not. This resembles a waterfall way of working.

Scrum is immutable. The essence of Scrum is lost when Scrum is customized to include a Sprint Zero.

The inability of Scrum Teams to get the development work started immediately indicates weaknesses in the way that organization works today. When Scrum is newly applied, it makes the existing weaknesses in the organization transparent. That is the strength of Scrum and organizations should leverage the transparency that Scrum brings in. Organizations should commit to improve the identified weaknesses. Only then will the truly self-managing teams emerge.

Making exceptions to weaknesses, in this case Sprint Zero, will hide the opportunities for improvements. The status quo will continue.

Another example is that some teams will create an exclusive Sprint called the "hardening Sprint" to enhance the technical quality of the Product Increment to make it production ready. Instead of looking at the situation by asking the right question "Why is the team not able to produce a production-quality Product Increment in the original Sprint itself?" this weakness is hidden under disguise of false Scrum terminology such as "hardening Sprint."

The Ongoing Act of Product Backlog Refinement – Starts before the first Sprint and continues

Purpose: The Product Owner and Developers continuously refine the Product Backlog Items. The items are refined until they are transparent enough for the Developers to estimate and confirm that they can be Done within a Sprint. When a Product Backlog Item reaches this level of transparency, it is also known as "Ready." The Developers are responsible for all estimates. The Product Owner may influence the Developers by helping them to understand and select trade-offs, but the people who will perform the work make the final estimate.

The Definition of Ready is a shared understanding by the Product Owner and the Developers Team regarding how much clarity the Product Backlog Items should have before they are introduced at Sprint Planning.

A Scrum Team should not wait until Sprint Planning to refine the items in a "just in time" manner. It is good practice to have enough items in a "Ready" state for at least one or two upcoming Sprints.

When: Backlog Refinement starts before the first Sprint and goes on throughout all Sprints. The Scrum Team decides how and when refinement is done. However, Product Backlog Items can be updated at any time by the Product Owner or at the Product Owner's discretion.

Timeline: This is probably the only activity in the Sprint where the Developers perform work outside the current Sprint Goal. Because it is important to ensure that there will always be some "Ready" items that can be selected for the next Sprint.

But care should be taken to optimize the time which the team spends on this act. Though this is not a time-boxed activity,

Refinement usually consumes no more than 10% of the capacity of the Scrum Team.

What is achieved: Product Backlog Items are decomposed by analyzing their intent. Each item is added with a description, order, estimate, and value. Higher-order Product Backlog Items are usually clearer and more detailed than lower-order ones. More precise estimates are made based on the greater clarity and increased detail, the lower the order, the less detail.

-------------------Question- 5--------------------------

In Backlog Refinement sessions, the Developers perform development activities such as coding and testing.

a) True

b) False

-------Answer-------

The Developers help in refining the items to a level such that the team can complete the items to Done status within a Sprint. They do this by analyzing, putting together solutions/designs, decomposing, and adding details, but there is no technical development activity during this refining session. Correct answer is 'b.'

-------------------Question- 5--------------------------

--------------------------DE-TOUR--------------------------

Spike: During the Product Backlog Refinement, a Team can perform technical analysis or design to understand the work involved. In situations where there is less clarity to estimate the effort and in-depth technical analysis or development (coding) is needed, the team can go for a "Spike" with the consensus of the Product Owner. A Spike is a time-boxed research activity to prove or disprove something and gain more clarity. Note that the time for Backlog Refinement including any Spike should not exceed 10% of the Scrum Team's time. The term 'Spike' is not part of the Scrum definition.

--------------------------DE-TOUR--------------------------

Backlog Refinement produces: Product Backlog

Content: A Product Backlog contains Product Goal and an ordered list of Product Backlog Items, each one having a set of attributes. Attributes often vary with the domain of work. In general, each Item at least has a description, order, and estimate. The Product Backlog Items together represent all that is needed in the product. The items include features, functions, requirements, enhancements, and fixes that constitute the changes to be made to the product in future releases. It is the single source of requirements for any changes to be made to the product.

Product Goal: It is a commitment attached to the Product Backlog. The Product Goal describes a future state of the product which can serve as a target for the Scrum Team to plan against. The Product Goal is in the Product Backlog. The rest of the Product Backlog emerges to define "what" will fulfill the Product Goal.

A product is a vehicle to deliver value. It has a clear boundary, known stakeholders, well-defined users, or customers. A product could be a service, a physical product, or something more abstract.

The Product Goal is the long-term objective for the Scrum Team. They must fulfill (or abandon) one objective before taking on the next.

Purpose: To maximize the transparency of what is required for the product, the order the team will work on next, and the estimate of the work involved.

Owner: The Product Owner is responsible for the Product Backlog, including its content, availability, and ordering.

Format: Nothing specific, as attributes often vary with the domain of work. Normally, each item at least has a description, order, and estimate. Usually the higher-order Product Backlog Items are clearer and more detailed than lower-order ones.

Lifecycle: A Product Backlog is never complete. The earliest development of it only lays out the initially known and best-understood requirements. The Product Backlog evolves as the product and the environment in which it will be used evolves. The Product Backlog is dynamic; it constantly changes to identify what the product needs in order to be appropriate, competitive, and useful. As long as a product exists, its Product Backlog also exists. As a product is used and gains value and the marketplace provides feedback, the Product Backlog becomes a larger and more exhaustive list. Requirements never stop changing, so a Product Backlog is a living artifact. Changes in business requirements, market conditions, or technology may cause changes to the Product Backlog.

Fig. 12 shows a sample Product Backlog. Remember that there is no standard format for the Product Backlog.

Product Backlog Item	Value (value point)	Effort (point)	Order
A registered customer can place request online to increase credit limit	50	21	Critical - 2
A business operations officer can get a report of credit increase limit applications by channel for last one month, in pdf format	25	8	Important - 2
Fix the the bug that doesn't display the credit increase preapproval offer for the customers logging in <xx> browser	40	13	Critical - 1
A registered customer can opt to provide a contact number and convenient time, to know if there are other ways to get more credit	5	8	Desired - 1
A registered customer can place request to increase credit limit from a smartphone with <xx> O/S	30	13	Important - 1

Fig 12- Sample of Product Backlog

------------------Question- 6-a-----------------------

Which is the correct statement?

a) The Product Backlog cannot be changed without a change request to the Product Owner.

b) The Product Backlog is not updated when a Sprint is in progress. Changes to team size may cause changes in the Product Backlog.

c) The Product Backlog can be updated anytime by the Product Owner. Changes in business requirements, market conditions, or technology may cause changes to the Product Backlog.

-------Answer-------

Correct answer is 'c.'

------------------Question- 6-a-----------------------

Defining the Definition of Done, which can provide clarity on the required work standard

We have already discussed how the Definition of Done provides a common understanding to the Scrum Team members about what it means for the work to be complete. In addition to being a standard for measuring completion, the Definition of Done is a reminder to the Developers about the need to account for all the end-to-end work required to meet the conditions. Teams can use this information to find out how many items they can select during Sprint Planning.

Who defines the Definition of Done?

If the Scrum Team is going to work on an existing Product or System, there should be an existing Definition of Done that is a standard for any work performed on this Product or System. The team should start with that.

If the Definition of Done for an Increment is part of a development organization's conventions, standards, or guidelines, all Scrum Teams must follow that as a minimum.

If there is no existing Definition of Done, the Scrum Team must define the Definition of Done as appropriate for the product.

As Scrum Teams mature, it is expected that their Definitions of Done will expand to include more stringent criteria for higher quality.

What if multiple Scrum Teams are working on the same Product?

If there are multiple Scrum Teams working on a system or product release, the Developers on all the Scrum Teams must mutually define the Definition of Done. Each Increment is additive to all prior Increments and thoroughly tested, ensuring that all Increments work together. It is not necessary for the Definition of Done to be the same but a mutually defined Definition of Done should enable the combined Increments to be valuable and useful.

Should the Product Owner approve the Definition of Done?

It is essential for the entire Scrum Team including the Product Owner to be well aware of the definition. Though there is no need for formal approval from the Product Owner, Scrum Team should define it with Product Owner's input.

While the Product Owner needs to be involved and made to understand the conditions, it is the Developers' responsibility to define the conditions in a verifiable way because many of these conditions are usually about technical quality. For example, the Product Owner may want a condition such as "The Increment should be thoroughly tested because it will be released to production," and the Developers may define it as "The Increment should pass all the automated unit tests with 95% code coverage."

-------------------Question- 6-b------------------------

How does a Scrum Team define the Definition of Done?

a) It always creates a new Definition of Done.
b) It checks with the Product Owner if a Definition of Done is required and then proceeds to create one.
c) None of the above.

-------Answer-------

If there is an existing Definition of Done (that is the organizational standard), Scrum Team must use it. Only when it doesn't exist, Scrum Team must create a Definition of Done. Correct answer is 'c.'

-------------------Question- 6-b------------------------

Can new teams define a relatively easy Definition of Done?

The purpose of each Sprint is to deliver Increments of valuable and useful functionality that adhere to the Scrum Team's current Definition of Done. Developers deliver an Increment with product functionality every Sprint. This Increment is useable, so the Product Owner may choose to immediately release it.

The Definition of Done should not be set with diluted quality measures with the objective of making it easy to meet. Unless the Increment is of useful quality, the Scrum Team cannot get feedback from actual usage. Diluting the Definition of Done will hide the current weaknesses in Product Development.

Given this, even a new team should define it such that the Increment will be production fit. However, the definition should contain conditions that are realistic to motivate the team. Then it can be continually improved by the maturing team's ability to perform all that is required. Having a realistic Definition of Done for a new team means that the working Increment may have known bugs, but they are transparent between the Developers and the Product Owner.

The total budget and number of Sprints are decided in which event?

Parameters like timelines and budget are reviewed in the Sprint Review. Work standards like Definition of Done are reviewed in the Sprint Retrospective. However, Scrum does not clarify when these parameters are decided for the first time. Unless otherwise inferred through the question, it is safe to assume that they are defined before the first Sprint at the time when just enough of a Product Backlog is put together.

Rules:

- Only the Product Owner can finalize what can be added to the Product Backlog.

- If multiple Scrum Teams work together on the same product, only one Product Backlog is used to describe the upcoming work on the product. An attribute that groups the items by each team may then be employed.

- Estimating Product Backlog Items: Only the people that will do the development work on the Product Backlog Items, i.e. the Developers, can finalize the estimates. Estimating the Product Backlog Items is a continuous process through the ongoing act of Backlog Refinement. This estimation is done by the Developers after the Product Owner's input.

- Measuring the value of the Product Backlog Items: It is mandatory for the Product Owner to assign a value for each Product Backlog Item. However, Scrum does not prescribe any technique to measure the business value of the Product Backlog Items. The Product Owner may use one or more appropriate techniques in the industry to estimate the business value.

- Ordering the Product Backlog Items – based on collective value: The Product Backlog Items must be ordered. However, though it is generally ordered based on the relative value of individual items, it is not necessarily a hard rule. It is based on the criteria which are deemed most appropriate by the Product Owner in order to maximize the collective value of the Product and the Scrum Team's work. In addition to ordering by the business values, the Product Owner can consider other factors like cost, risk, coherence, organizational policies, and also can seek the input of the Developers about the technical dependencies between the Product Backlog Items. The Product Owner shall try to optimize the ordering such that it eventually leads to the product Increment with the best overall value that is both releasable and useable.

------------------Question- 7-------------------------

Which of the following statements is not correct?

a) Only the people who perform the work can finalize the estimate of Product Backlog Items.

b) The Product Owner always orders the Product Backlog Items based solely on the value of each individual item compared to another item.

c) Multiple Scrum Teams working on the same product should have only one common Product Backlog.

d) A Scrum Master can author a Product Backlog Item for the Product Owner's consideration.

e) The Developers finalize all estimates.

-------Answer-------

The Product Owner is the ultimate authority of finalizing what needs to be added to the Product Backlog. However, they can have others provide them with suggestions. So, a Scrum Master can always author an item for the Product Owner's consideration. The Product Owner strives to maximize the collective value of the Product and the Scrum Team's work. To achieve that, they can choose to follow any appropriate logic for ordering. It need not always be the individual business value. Correct answer is 'b.'

------------------Question- 7-------------------------

Ordering the Product Backlog Items – Practice Questions

The logic behind which parameter is chosen to order the Product Backlog is bit tricky. Go through the following questions to understand the best choice for a given context.

------------------Question- 8-------------------------

The Product Backlog is ordered by

a) The individual Product Backlog Item's value.
b) Whatever increases the overall value of the team's work.
c) Whatever is deemed as appropriate by the Product Owner.

-------Answer-------

A clear answer is there. Correct answer is 'c.'

------------------Question- 8-------------------------

------------------Question- 9-------------------------

The Product Backlog is ordered by

a) The individual Product Backlog Item's value.
b) Whatever increases the overall value of the team's work.
c) The priority of senior management.

-------Answer-------

There is no clear answer. Choose the next best choice. Correct answer is 'b.'

------------------Question- 9-------------------------

------------------Question- 10-------------------------

The Product Backlog is ordered by

a) The individual Product Backlog Item's value.
b) Whatever is deemed appropriate by technical and domain experts.
c) The priority of senior management.

-------Answer-------

There is no clear answer. Choose the next best choice. Correct answer is 'a.'

------------------Question- 10-------------------------

Summary

- Scrum starts with just enough preparation and pursues a "value discovery and maximization" journey.
- The journey is concluded when enough value is delivered or when the investment becomes unjustified or when allocated resources are exhausted.
- The Sprint is the heart of Scrum. It is like a mini project.
- A Sprint enables predictability by ensuring inspection and adaptation of progress towards a Sprint Goal at least every calendar month.
- Each Sprint has a definition of what (is to be built), why (Sprint Goal), how (a design and flexible plan that will guide building it), the work, and the resultant product.
- Unlike the traditional approach of bringing people to the work, Scrum brings the work to the people. It is staffed with a constant team which avoids staffing complexities.
- Unlike the traditional approach of big upfront design, the Scrum Teams evolve the technical design throughout the journey.
- Unlike the traditional approach of an upfront complete requirements specification, a Scrum Team starts the first Sprint with just enough Product Backlog.
- The Product Backlog is an ordered list of everything that may be needed in the product and is the single source of requirements for any changes to be made to the product.
- The first Sprint needs just enough Product Backlog that only lays out the initially known and best-understood requirements. The Product Backlog is a living artifact that evolves and constantly changes to identify what the product needs in order to be appropriate, competitive, and useful. As long as a product exists, its Product Backlog also exists.
- Product Backlog refinement is the ongoing act of the Product Owner and the Developers collaboratively adding detail, estimates, order, and value to items in the Product Backlog.
- Refinement starts before the first Sprint and goes on through all the Sprints, but consuming no more than 10% of the capacity of the Scrum Team.
- The Product Owner orders the Product Backlog Items by whatever method they think is the most appropriate to maximize the value of the product.
- Only the Developers can finalize the estimates of Product Backlog Items.
- A Sprint starts with the Sprint Planning event.
- The input to this meeting is the Product Backlog, latest product Increment if any, projected capacity of the Scrum Team during the Sprint, and past performance of the Scrum Team if available.
- If there is an existing Definition of Done (that is the organizational standard), Scrum Team must use it. If it doesn't exist, Scrum Team must create a Definition of Done.

#

Chapter 2 – Staffing roles with the right skills

--------------------------DE-TOUR--------------------------

In this chapter, the following are included for clarity and context. They are not part of The Scrum Guide:
Reporting to Management, Same person playing Scrum Master and Product Owner roles

--------------------------DE-TOUR--------------------------

There are only three roles in Scrum. Before starting the first Sprint, all these roles must be staffed.

There are external stakeholders like senior management, customer, user, etc. Also, there are technical and domain experts that provide insights and input to the Scrum Team. However, those are the people outside (not within) the Scrum Team.

What is their accountability?

Scrum Master	Product Owner	Developers
Establish Scrum as defined in the Scrum Guide. Help teams and organization understand and enact it by spirit. Help improve teams' practices	Maximize value from Scrum Team's work. Keep the team and stakeholders communicated about the Product Goal and Product Backlog	Self-manage and deliver any aspect of usable Increment every Sprint. Improve the way of working Sprint by Sprint

What do they manage?

Scrum Master	Product Owner	Developers
Not people. Manage the Scrum implementation	Not people. Manage the Product Backlog	Not people. Manage the Sprint Backlog and their development work

Who do they work with?

Scrum Master	Product Owner	Developers
One or more Scrum Teams	One or more Scrum Teams	One Scrum Team dedicated

Can one person share two accountabilities?

Scrum Master	Product Owner	Developers
In addition to being a Scrum Master, they can also actively work on Sprint Backlog items as Developers	In addition to being a Product Owner, they can also actively work on Sprint Backlog items as Developers	You already saw that Developers can also be a Scrum Master or Product Owner (who chose to actively work on Sprint Backlog and hence count as Developers as well)

Who do they report to?

Scrum Master	Product Owner	Developers
With respect to their (Scrum Team's) work, they self-manage it (they internally decide who does what, when, and how). With respect to Human Resource aspects, they report to some authority or management layer in the organization		

-------------------------DE-TOUR-------------------------

Scrum does not directly clarify if a Product Owner can also be a Scrum Master.
Let us consider the primary job of each of these roles. A Product Owner wants to maximize/optimize the value of the Scrum Team's work. It is possible for them to push for "more work" while compromising the Definition of Done or technical quality requirements. They may also overlook the Developers' empowerment of deciding for themselves how much work they can select for a Sprint.
In such conflicting situations, a Scrum Master teaches both the Product Owner and the Developers about their empowerment and balances.
So, we can see that there could be situations involving conflict of interest if one person plays both roles.
Let us examine this further.
Think about the fundamentals of Scrum. It is a risk reduction framework for building complex products. The Scrum Team is not only self-sufficient. It is also balanced with respect to accountabilities. The risks and subjectivities associated with a traditional one-man centric Project are mitigated by distributing the accountabilities between the three roles. Now, what happens if you bring the Scrum Master and Product Owner under a single hat? You can sense the concentration of responsibilities. Will it be Scrum? The answer is – It increases risk and reduces self-management thus undermining Scrum. So, the Product Owner and the Scrum Master are separate accountabilities played by different individuals.

-------------------------DE-TOUR-------------------------

Summary

This chapter is too short to review. Reread, if necessary.

#

Chapter 2.1 - Scrum Master - Scrum Guardian and Servant-Leader

What is expected from the Scrum Master?

- Scrum Coach: The Scrum Master coaches the Scrum Team and ensures that it understands and adheres to the Scrum way of working that include Scrum theory, practices, rules, and values. As needed, the Scrum Master facilitates the events for the Product Owner and the Developers and ensures timeboxing.

- Servant Leadership: Instead of commanding and controlling, the Scrum Master has a strong inclination to serve the team by helping them to realize their own potential. The Scrum Master facilitates the team to become self-managing by coaching them to come up with a work plan and make daily decisions by themselves.

- Removes Impediments: While the Scrum Master mentors the team to solve issues by themselves, they own and remove other impediments that are outside the influence of the Team. In organizations starting Scrum, the Scrum Teams may get a lot of external interruptions and requests for additional tasks. The Scrum Master helps those outside the Scrum Team understand which of their interactions with the Scrum Team are helpful and which aren't. The Scrum Master helps everyone change these interactions to maximize the value created by the Scrum Team.

- Champion of Transparency: The Scrum Master helps the team to increase transparency of planned work, actual progress, and impediments. The Scrum Master spends time detecting incomplete transparency by inspecting the artifacts, sensing patterns, listening closely to what is being said and not being said, and detecting differences between expected and real results.

-------------------Question- 11--------------------------

Which of the following statements is true?

a) Scrum Master is a management position. A person with strong project management experience in delivering results is a good fit.

b) Scrum Master is an optional position. An alternative is to train the team on Scrum before they start, and they can self-manage without a Scrum Master.

c) Scrum Master is a management position. A person with Scrum experience and a coaching style of servant leadership is a good fit.

d) None of the above.

-------Answer-------

The Scrum Master is required to have Scrum experience and a strong inclination towards a Servant Leadership style. It is a mandatory position to keep the Scrum Team and the organization focused on Scrum. Although the Scrum Master does not directly manage people, the role is a management position because the Scrum Master manages the Scrum framework implementation. Correct answer is 'c.'

-------------------Question- 11--------------------------

Additional services to the Developers:

- The Scrum Team can expect the Scrum Master to coach them to become a cross-functional team and help them create high-value products. They can also expect the Scrum Master to coach them on how to handle organizational forces in less mature Scrum environments.

- The Scrum Team can expect the Scrum Master's help, whenever required, to find the best sub-techniques to accomplish their work within the Scrum framework. An example of a sub-technique is using a burn-down graphic to track progress of product completion and Sprint Backlog completion.

Additional services to the Product Owner

- Finding techniques for effective Product Backlog management.
- Helping the Scrum Team understand the need for clear and concise Product Backlog Items.
- Understanding product planning in an empirical environment.
- Ensuring the Product Owner knows how to arrange the Product Backlog to maximize value.
- A non-cooperative Product Owner is an impediment to be resolved by the Scrum Master.
- If a parallel or competing authority to the Product Owner directs or interferes with the team's work, then this is an impediment that needs to be resolved by the Scrum Master. As a first step, the Scrum Master may need to coach the Product Owner about the sole authority they only have in deciding what the team should work.
- Ensuring that goals, scope, and product domain are understood by everyone on the Scrum Team.

-------------------Question- 12-a-------------------------

Select all that apply. What are some examples of Product Backlog management techniques where a Scrum Master can coach the Product Owner and the Developers?

a) Creating a common standard that defines the preferred level of description and transparency each Product Backlog Item should meet before introducing them in Sprint Planning. The Team can then use this standard as a guideline to decompose the Items.

b) In addition to using value, a Product Owner can choose input from the Developers on ordering the items based on their technical coherence.

c) Choosing a tool to manage the Product Backlog.

d) Techniques like writing the items in the form of user stories and their Acceptance Tests.

-------Answer-------

A Scrum Master coaches the Product Owner and the Developers about managing the Product Backlog to facilitate empiricism-based product planning and arranging the items so that the order can maximize overall value. The Scrum Master also coaches the Product Owner to collaborate with the Developers on ordering. Correct answers are 'a,' 'b,' and 'd.'

-------------------Question- 12-a-------------------------

-------------------Question- 12-b-------------------------

The Developers in a Scrum Team do not have enough understanding of the domain of the Product they develop. The Scrum Master need not take an active interest in improving this issue as it is not related to Scrum.

a) True.

b) False.

-------Answer-------

A Scrum Master coaches the Developers to create high-value products. Part of the coaching, the Scrum Master ensures that goals, scope, and product domain are understood by everyone on the Scrum Team. Correct answer is 'b.'

-------------------Question- 12-b-------------------------

Services to the Organization

- The Scrum Master coaches Scrum to the stakeholders on how they should interact with the Scrum Team in order to maximize the value of the product created.

- The organization can expect the Scrum Master to coach the employees about Scrum and empirical product development and self-management to cause organizational change.

- Further it can expect the Scrum Master to help plan Scrum implementations, collaborate with stakeholders and other fellow Scrum Masters to drive the organization-wide Scrum adoption, and increase the productivity of the Scrum Teams.

------------------Question- 13-------------------------

The role of the Scrum Master with respect to the Scrum artifacts is to:

a) Coach the Team to increase the transparency of the artifacts.

b) Decide the format of the artifacts and ensure that the Team follows it.

c) Own the artifacts and be responsible for keeping them up to date.

-------Answer-------

Correct answer is 'a.'

------------------Question- 13-------------------------

Rules

The Scrum Master need not be a technical member of the Scrum Team. They can be full-time or part-time for a Scrum Team. They are usually a Scrum Master for more than one Scrum Team in parallel. It is mandatory for the Scrum Master to participate in all events except the Daily Scrum.

------------------Question- 14-------------------------

In a Scrum Team, if the Scrum Master is also a Developer, then

a) They should not participate in the Daily Scrum.

b) They must participate in the Daily Scrum.

c) They can participate in the Daily Scrum only if the Developers invite them.

-------Answer-------

Every Developer must participate in all the events of Scrum. In this case, the Scrum Master is also a Developer. Correct answer is 'b.'

------------------Question- 14-------------------------

Summary

- The Scrum Master is responsible for promoting and supporting Scrum as defined in the Scrum Guide. Scrum Masters do this by helping everyone understand Scrum theory, practices, rules, and values.

- The Scrum Master is a servant-leader for the Scrum Team. The Scrum Master helps those outside the Scrum Team understand which of their interactions with the Scrum Team are helpful and which aren't. The Scrum Master helps everyone change these interactions to maximize the value created by the Scrum Team.

- The Scrum Master also removes impediments and is the champion of transparency.

- They facilitate Scrum events as requested or needed.

- They serve the Product Owner in finding techniques for effective Product Backlog management, understanding product planning in an empirical environment, and knowing how to arrange the Product Backlog to maximize value. The Scrum Master ensures that goals, scope, and product domain are understood by everyone on the Scrum Team as well as possible.

- They serve the Scrum Team in coaching self-management and cross-functionality, helping to create high-value products, and removing impediments to the Scrum Team's progress.

- They serve the organization in its Scrum adoption, planning Scrum implementations within the organization, helping employees and stakeholders understand and enact Scrum and empirical product development, causing change that increases the productivity of the Scrum Team, and working with other Scrum Masters to increase the effectiveness of the application of Scrum in the organization.

#

Chapter 2.2 - Product Owner - Rigorous value maximization

--------------------------DE-TOUR--------------------------

In this chapter, the following are included for clarity and context. They are not part of The Scrum Guide:
Product Owner in contract work, Issues in implementing the Product Owner role

--------------------------DE-TOUR--------------------------

What is expected from the Product Owner?

Collaborative Value Maximization: A strong inclination to maximize the value of the product by constant collaboration in developing the Product Goal, defining the product features that will meet this goal, keeping them ordered, and feeding back the insights into product refinement. They bring transparency to the above activities by owning and maintaining the Product Backlog artifact:

- Responsible for developing and explicitly communicating the Product Goal.
- Responsible for maintaining the Product Backlog order by sequencing the Items to best achieve goals and missions.
- Responsible for maintaining the Product Backlog content by clearly expressing Product Backlog Items and by updating it with the latest insights and customer or market needs.
- Responsible for maintaining the Product Backlog availability by ensuring that the Product Backlog is visible, transparent, and clear to all, and shows what the Scrum Team will work on next.

Product Owners Optimize the Value of the Scrum Team's Work by ensuring the Developers understand the Product Goal and the Items in the Product Backlog to the level needed.

Product Owners Increase Transparency of what product features are completed so far and what the Teams are going to work on next. They use the Sprint Review to update this to stakeholders and also forecast a product completion date.

-------------------Question- 15--------------------------

During a Sprint Review, the stakeholders notice that the product development progress is not clearly visible and lacks transparency. Moreover, they are not able to understand the Team's next steps. Who bears the primary responsibility for this status?

a) Scrum Team
b) Scrum Master
c) Product Owner
d) Developers

-------Answer-------

The entire Scrum Team is responsible for "how they plan and perform their work." So, if there is a question about who is responsible for the failure or success of the Scrum work, the answer is absolutely the Scrum Team.

The Scrum Master of course needs to help the Product Owner in coaching techniques for better Product Backlog management, increasing the transparency of the Backlog, and more. So, the Scrum Master's responsibility is to coach the team to adhere to Scrum and its principles. However, the Scrum Master does not bear the primary responsibility for those items that are clearly owned by a specific role.

There are clearly defined accountabilities for each role. In this case, the question is - who is accountable for a specific activity of Backlog management including its transparency. The following are guidance from The Scrum Guide:

- *The Product Owner remains accountable for Product Backlog Management.*
- *The Product Owner is also accountable for ensuring that the Product Backlog is transparent, visible, and understood.*

So, the best choice is the "Product Owner."

Correct answer is 'c.'

------------------Question- 15-------------------------

What could go wrong with the implementation of the Product Owner role?

It is common for an organization to identify an existing business manager and staff them in the Product Owner role. While this is perfectly acceptable, the Product Owner role is significantly more empowered. This role also requires strong inclination towards a collaborative style of product development.

- <u>They are the Owners for evolving the Product and should be allowed to act accordingly</u>: The Product Owner represents the desires of the organization in the Product Backlog. The Product Owner's decisions are visible in the content and ordering of the Product Backlog. For the Product Owner to succeed, the entire organization must respect their decisions. Those wanting to change a Product Backlog Item's priority must address the Product Owner. The Product Owner is the only point of contact for the stakeholder, customer, and sponsor.

- <u>They should be willing to be continuously available to the Team</u>: They continually work with the Team to make them understand the product needs and derive the best value out of their work.

- <u>They are the ultimate authority of the Team's work and should act accordingly</u>: The Product Owner leverages the team's skills to continually shape the product and uncover new knowledge. So, it is essential that the Team should only be working on their product's needs. The Team should not be "switching the context" of their work by external authorities directing them to do "some other" work. So, the Product Owner is the ultimate authority on what the Team should work next. No one is allowed to tell the Developers to work from a different set of requirements, and the Developers are not allowed to act on what anyone else says. Even the CEO of the organization cannot request the team to work on something else. Anybody wanting to change the priority must address the Product Owner.

------------------Question- 16-------------------------

A Developer is requested by an important stakeholder to help them with some external task because it is urgently required by the organization's board. The Developer referred them to Product Owner. In this case, the Scrum Master

a) Should do nothing, since the Developer's action was correct.
b) Should coach the Developer to support senior management requirements.
c) Should form a sub-team that can take up such external requests.

-------Answer-------

No one is allowed to tell the Developers to work from a different set of requirements, and the Developers are not allowed to act on what anyone else says. Correct answer is 'a.'

------------------Question- 16-------------------------

Who should play the Product Owner role in contract efforts?

A Product Owner is not necessarily the customer. If the project is a contract work performed by a service organization, a person within the service organization can play the Product Owner role and represent the customer.

Rules

- They can be full or part time for a Scrum Team. They are usually a Product Owner for more than one Scrum Team in parallel.

- The Product Owner is accountable for Creating Product Goal, Managing Product Backlog, and Making them visible and transparent to the stakeholders.

- The Product Owner can delegate one or more responsibilities to others in the Team, but they are still accountable for the product value.

- It is mandatory for the Product Owner to participate in all events except the Daily Scrum.

- No one can change the Product Backlog other than the Product Owner. However, the Developers, Scrum Master, and stakeholders can recommend items that could be added to the Product Backlog.

- No one can cancel the Sprint other than the Product Owner. But the Developers, Scrum Master, and stakeholders can influence the Product Owner to make that decision.

- If a stakeholder or customer needs to communicate anything to the Team, they should direct such communications through the Product Owner.

-------------------Question- 17-------------------------

The number one priority of the Product Owner is

a) Managing the development work.
b) Guarding the Scrum Team from any interruptions.
c) Maximizing the value of the Scrum Team's work.
d) Testing the Development Developers' work against detailed requirements.

-------Answer-------

The Product Owner does not manage or direct the development work during the Sprints. It is the responsibility of the Scrum Master to manage external interruptions. While the Product Owner may help with testing, they are not responsible for detailed testing. Correct answer is 'c.'

-------------------Question- 17-------------------------

Summary

- The Product Owner is responsible for maximizing the value of the product and the work of the Scrum Team.

- The Product Owner is the sole person responsible for creating the Product Goal, managing the Product Backlog, and keeping the team updated about the Product Backlog. The Product Owner may have the Developers do some of the Product Backlog management activities. However, the Product Owner remains accountable.

- No one including the CEO can modify the Product Backlog. Those wanting to change a Product Backlog Item's priority must address the Product Owner.

#

Chapter 2.3 - Developers - Engine that converts needs into values

What is expected from the Developers?

- Developers are professionals that "perform" the end-to-end development work of delivering a valuable and usable Increment of Done product at the end of each Sprint.

- Developers work full-time within a Scrum Team. They are small enough to reduce communication complexities and big enough to include the required skills to perform a complete work.

Optimal Scrum Team size is small enough to remain nimble and large enough to complete significant work within a Sprint. Fewer Developers in Scrum Teams decreases interaction and results in smaller productivity gains. Such teams may encounter skill constraints during the Sprint causing the Developers to be unable to deliver a valuable and useful Increment. Having more than ten members requires too much coordination. Large Scrum Teams generate too much complexity for an empirical process to manage. Typical Scrum Teams contain 10 or fewer people including Product Owner and Scrum Master.

-------------------Question- 18--------------------------

An organization forms Scrum Teams each with a size of 15 members because it is a convenient way to map their current teams that are of the same size.

a) This is okay. It will speed up the Scrum transformation.

b) Scrum is immutable. It is recommended that the Scrum Team follow the guideline of ten or fewer people in order to drive the Scrum-based change.

c) It is up to the Teams to decide how they want to be formed.

-------Answer-------

Scrum recommends ten or fewer people, which is small enough to reduce complexity but big enough to have the required skills and capacity. Anything less may get only marginal productivity gains. Anything more will invite complexity. Correct answer is 'b.'

-------------------Question- 18--------------------------

- The team is SELF-MANAGING. The team is empowered in self-management in such a way that no one, including the Product Owner, Scrum Master, or any other manager external to the Scrum Team directs or commands them on how to perform their work. The team self-manages by focusing on the goal and working within the time-box.

-------------------Question- 19--------------------------

A Scrum Team gets into a situation where a conflicting Developer's behavior causes issues. Who is responsible for removing this issue?

a) Management

b) Product Owner

c) Scrum Master

d) Scrum Team

-------Answer-------

Think about who is responsible for identifying and removing different types of issues. The Scrum Master is responsible for removing impediments outside the Scrum Team's influence. Also, they are responsible for causing change that increases the productivity of the Scrum Team.

In this case, the issue faced by the Scrum Team is well within the influence of the Scrum Team to resolve. So, the Scrum Master should coach the team to resolve such items themselves. If the Scrum Master actively takes steps such as removing this person from the Team, it will lead to a diminished inclination of the Scrum Team to resolve internal problems for themselves in the long run. Correct answer is 'd.'

-------------------Question- 19-------------------------

- The team is CROSS-FUNCTIONAL. The team is unified in such a way that there are no specialist roles or sub-teams.

Individual Scrum Team members may have specialized skills and areas of focus, but accountability belongs to the Scrum Team as a whole. The Sprint Goal binds the Team together. Scrum recognizes no titles for the Developers other than Developer, regardless of the work being performed by the person.

Scrum recognizes no sub-teams in the Scrum Team, regardless of particular domains that need to be addressed like testing or business analysis.

-------------------Question- 20-------------------------

A Scrum Team has many Developers with cross-functional skills. This team builds the product Increment that almost meets the conditions set by the Definition of Done. To completely meet the Definition of Done, they hand over the product Increment to another team outside the Scrum Team for specialized testing. Is this truly a cross-functional team?

a) Yes

b) No

-------Answer-------

A Scrum Team should possess all the skills needed to deliver a valuable and useful product that meets the Definition of Done without external help. Correct answer is 'b.'

-------------------Question- 20-------------------------

The resulting synergy of this structure, empowerment, and unification optimizes the Scrum Team's overall efficiency and effectiveness. Without any management or direction from outside, they strive to balance flexibility, creativity, and productivity, so they can maximize the value of their work.

An individual can be cross-skilled, but it is not mandatory

A Scrum Team may contain various specialists needed to achieve the Sprint Goal. For example, there could be Programmers, Testers, UI modelers, Architects, Technical writers, etc. But there is NO special name for any of them. Irrespective of their field of specialization, each one of them (other than the Product Owner and the Scrum Master) is called a Developer.

While the specialist should identify themselves as part of the team and learn additional skills to collectively deliver the Sprint Goal, there is no barrier to personally enhancing their vertical competencies and continue to specialize.

For example, an architect may be added to the team if the work requires that skill, but will not be given a formal role or title called "Architect." Though this team member will contribute to the architectural aspects of the effort, he or she along with the entire team is responsible for the progress of the collective Sprint Goal and is expected to help the team to reach that goal. In the process, this team member may expand their skillset to become cross-functional.

-------------------Question- 21--------------------------

A Scrum Team has technical specialists in addition to Developers. They perform their work when the Sprint Backlog needs their special skills, but they are idle otherwise.

a) Continue to have specialists deliver fully integrated Increments. Gradually facilitate the team to organize their work to fully leverage these specialty skills. If required, they can enhance everyone's domain of expertise, so that everyone is productive without idle time.

b) Let the Project Manager coordinate their staffing needs and plan partial allocations to avoid idle time.

c) Defer and accumulate the special work to later Sprints until it needs full-time specialists. Add them to the team for that timeframe alone. Prior to that, deliver the Increment with stubbing.

-------Answer-------

Correct answer is 'a.'

-------------------Question- 21--------------------------

The Scrum Team is self-managing. Does it mean there are no managers?

The fact that the Scrum Team is self-managing does not mean that there are no managers for team members. But they may be "people aspect" managers and not "product development work aspect" managers. These two types of management are explained below:

- Management of how the development work is planned, performed, and controlled: This is performed by the Scrum Team. No one outside the Scrum Team can command or direct the development work.

- Management of the people aspects of the Scrum Team: Every Scrum Team usually works within a larger organizational ecosystem. The people aspects of the Scrum Team, such as human resource management functions of hiring, firing, compensation, and other legal aspects are handled by an appropriate human resource authority in the organization.

-------------------Question- 22--------------------------

A Scrum Team decides that the frequency of the Daily Scrum should be reduced to once a week.

a) Such decisions need to be approved by the Team Manager.

b) Such decisions need to be approved by the Agile Coach.

c) The Scrum Team is self-managing. They can choose their practices.

d) The Scrum Master should coach the team on the essentials of conducting Daily Scrums.

-------Answer-------

Though individuals may report to some management authority on the "people aspects," there is no exclusive manager for a Scrum Team. In Scrum, there is no role called Agile Coach.

Self-management is about the empowerment for the Scrum Team to decide who does what, when, and how. As a guardian of the Scrum implementation, the Scrum Master should coach the team on Scrum essentials. Correct answer is 'd.'

-------------------Question- 22--------------------------

Rules

- The Developers as whole and not the individual is accountable for Sprint progress.

- They are responsible only for items related to the Sprint Goal and Sprint Backlog.

- During Sprint Planning, no one other than the Developers makes decisions about how many Backlog Items they will include in a Sprint.

- The Developers completely owns the Sprint Backlog. Nobody can change the Sprint Backlog other than the Developers. The Product Owner can influence the changes only with the consent of the Developers.

- It is mandatory for the Developers to participate in all Scrum events and the act of Backlog Refinement.

- There can be multiple Scrum Teams in parallel. If all of them work on the same product, there should be ONLY ONE common Product Backlog, and a mutually defined Definition of Done. A mutually defined Definition of Done enables their combined work outcome to be valuable and useful. Since there is only one Product Backlog, there should only be one Product Owner.

-------------------Question- 23--------------------------

In the middle of the Sprint, a Scrum Team finds that they have more room for additional work. They decide to change the Sprint Backlog by adding a few more Backlog Items from the Product Backlog. Who should be present to decide the additional work and accordingly modify the Sprint Backlog? Select all that apply.

a) Senior Developer of the Scrum Team

b) Scrum Master

c) All Developers

d) Product Owner

e) Scrum Team

-------Answer-------

Nobody can change the Sprint Backlog other than the Developers. So, they should be present. The Product Owner is responsible for optimizing the value of the Scrum Team's work and is needed to explain the content of the Product Backlog and give mutual consent on the next work. So, they also need to be present. Correct answers are 'c' and 'd.'

-------------------Question- 23--------------------------

Summary

- The Developers does the work of delivering a valuable and usable Increment of Done product at the end of each Sprint.

- Scrum Teams are structured and empowered by the organization to organize and manage their own work.

- Optimal Scrum Team size is small enough to remain nimble and large enough to complete significant work within a Sprint.

- Fewer Developers decreases interaction, increases skill constraints, and results in smaller productivity gains.

- More than ten members increases complexity and requires too much coordination.

- They are self-managing and cross-functional.

- Everyone (other than the Product Owner and the Scrum Master) is called Developers, regardless of the special work individuals may perform like testing, business analysis, etc.

- Each Developer may have a distinct work focus, but accountability belongs to the Developers as a whole.

- There are no sub-teams in the Scrum Team.

#

Chapter 3 - Interfacing with people outside Scrum

-------------------------DE-TOUR-------------------------

Very little information is found in Scrum about people outside the Scrum Team. This chapter provides some clarity.

-------------------------DE-TOUR-------------------------

Organization and Senior Management

This is not a Scrum role. This is an entity external to the Scrum Team, who sets the larger strategies of organizational change.

- The organization should have guidelines about how to structure the Scrum Teams, so it will empower these teams to organize and manage their work.

- If the organization does not yet understand the Scrum concept, it is the responsibility of the Scrum Master to coach the organization as well the Scrum Teams.

- The organization can expect the Scrum Master to coach the employees about Scrum and empirical product development, plan Scrum implementations, collaborate with other Scrum Masters for organization-wide Scrum adoption, and increase the productivity of Scrum Teams.

- The organization should understand the vital role played by the Product Owner and hence respect their decisions, so that the Product Owner can succeed in their role. The organization should provide the Product Owner with information that will help them increase the value of the product and capabilities.

------------------Question- 24-------------------------

The support of the organization or senior management

a) Is not needed for Scrum implementations.

b) Is not needed because there is no scope for management in Scrum.

c) Is needed to support the Product Owner to maximize the product value and the Scrum Master to coach and implement Scrum.

-------Answer-------

Correct answer is 'c.'

------------------Question- 24-------------------------

Stakeholder

Scrum has a specific meaning for the term 'stakeholder,' which refers to people such as customers, users, and all those who have the commonality of having a specific interest and knowledge in a product. The stakeholder is not a Scrum role. However, they influence the product development direction when **invited** by the Product Owner to participate in the Sprint Review.

Before starting the Sprints, it is highly useful to identify the stakeholders with whom the Scrum Team needs to work. There is no formal artifact to record such information. The stakeholder's information can be publicly displayed in the workplace of the Team.

- At any point in time, stakeholders have access to the information about the Scrum Team's progress towards a goal.

- During the Sprint Review, they collaborate with the Scrum Team on the next steps.

- They can influence the Product Owner to cancel a Sprint.
- They can approach the Scrum Master to understand Scrum and empirical product development.

How are the external people involved with Scrum Teams?

Organization Senior Management	Stakeholders	Technical/Domain Experts
There is no active involvement in Scrum activities. Scrum Master can approach for their support in implementing Scrum. Management can decide strategy for team structuring and empowering to encourage self-management	Can be invited by Product Owner to participate in Sprint Review	Can be invited by Scrum Team to participate in Sprint Planning

-------------------Question- 25-------------------------

Who must participate in the Sprint Review?

a) Scrum Team
b) Audit Team
c) Technical and domain experts
d) Stakeholders

-------Answer-------

Apart from the Scrum Team, stakeholders must participate only if they are invited by the Product Owner. Technical and domain experts are not stakeholders. If one of the options is 'invited stakeholders', it is a correct answer and needs to be selected along with 'Scrum Team.' Correct answer is 'a.'

-------------------Question- 25-------------------------

What can they expect from Scrum Teams?

Organization Senior Management	Stakeholders	Technical/Domain Experts
They can expect visibility of Product Backlog. They can expect Scrum Master's help to cause organizational change. However, they don't closely monitor or control the Scrum Teams	Required stakeholders can expect transparent information on progress towards the Product Goal	They can expect the Scrum Team to define the Definition of Done in compliance with common standards defined at organization level

-------------------Question- 26--------------------------

Select all that apply. Who performs inspections in Scrum events?

a) Product Owner

b) Corporate Audit Group

c) Developers

d) Senior Management

e) Technical Domain Experts

f) Invited Stakeholders

-------Answer-------

In Scrum, inspections are performed by those doing the work and those who have knowledge of the product. In addition to the Developers, the Product Owner and invited stakeholders inspect the product Increment during the Sprint Review. Correct answers are 'a,' 'c,' and 'f.'

-------------------Question- 26--------------------------

Summary

This chapter is too short to review. Reread, if necessary.

#

Quiz 2

1. You are working as a Scrum Master on a team. Your organization plans to conduct a road show on Scrum across the board. You are asked to contribute to some related events.

a) You will gladly volunteer because being a Scrum Master in an organization also involves the responsibility of coaching the organization.

b) You will politely refuse because you are working as a Scrum Master on a team that is fully focused on delivering the Sprint Goal.

c) As a leader, you will delegate this to some other team members so that they can benefit from the learning and visibility.

-------Answer-------

One of the Scrum Master's responsibilities is coaching Scrum to the organization. Correct answer is 'a.'

2. Since the Scrum Team is self-managing, it can create an additional role to represent the organization within Scrum.

a) True

b) False

-------Answer-------

Scrum has only three roles. The organization needs are best represented in the Product Backlog, which is available and visible to the required stakeholders in an organization. Correct answer is 'b.'

3. An important executive wants the Developers to include a highly critical feature in the current Sprint. The Developers

a) should work on that feature since an organization's priority is more important.

b) should ask the executive to work with the Product Owner.

c) should negotiate with the executive and identify an alternative backlog item of comparable scope in the current Sprint that may be removed to accommodate this critical feature.

-------Answer-------

The Developers should only work on the items related to the Sprint Goal. No one is allowed to tell the Developers to work from a different set of requirements, and the Developers aren't allowed to act on what anyone else says. Correct answer is 'b.'

4. The senior management or organization does not influence the Product Owner on how the product evolves in any way.

a) True

b) False

-------Answer-------

The senior management and the organization set the larger strategies of organizational change. Such strategies may influence the product development.

In addition, the organization enables the Product Owner to obtain information that will help them increase the value of the product and capabilities. When required, they directly engage the Product Owner in setting expectations of the product. So, though they do not play any active role within Scrum on product development, they influence it through other means. Correct answer is 'b.'

5. A Scrum Team must be cross-functional. It means

a) Each team member must be cross-skilled.

b) The team must have a mix of team members from each of the different technology functions of the organization.

c) The team must have all competencies needed to create the Increment according to the Definition of Done.

d) The team must divide their work according to subfunctions but sync up their work no later than the Sprint Review.

-------Answer-------

Cross-skilling of team members is desirable but not mandatory. Required skills are decided by what is needed per the Definition of Done and not by an organization's technology functions. The team works together every day not waiting for a last-minute sync up. Correct answer is 'c.'

6. A customer wants to communicate something very relevant and important about the product to the Scrum Team. Who should they talk to?

a) Since everyone on the team is accountable for product development, the customer should meet all of them together.

b) Product Owner

c) Scrum Master

d) Developers

-------Answer-------

The Product Owner may represent the desires of a committee or customer in the Product Backlog, but those wanting to change a Product Backlog item's priority must address the Product Owner. Correct answer is 'b.'

7. In a Scrum Team, only the Product Owner communicates with the stakeholders. There is no exception to this rule.

a) True

b) False

-------Answer-------

Stakeholders only communicate with the Product Owner with respect to Product Planning, its Progress, and the Product Backlog changes. In addition, the Scrum Master can work with the stakeholders to make them understand Scrum. Also, the entire Scrum Team collaborates with stakeholders during the Sprint Review. Correct answer is 'b.'

8. A Developer identifies a technical issue that requires other Developers to work together to solve this. Who needs to facilitate this?

a) Product Owner

b) Scrum Master

c) Developers

-------Answer-------

The Developers are responsible for performing development work and meeting the Sprint Goal. If they have any issue that is within their influence to solve, they are responsible for resolving it. The Scrum Master is responsible for impediments that are outside the Scrum Team's influence. Correct answer is 'c.'

9. In the early days of a Product Development's initial Sprint, the Product Backlog

a) Only lays out the initially known and best-understood requirements.

b) Is as comprehensive as required to complete the Product.

c) Is not shared with the Developers until it is supplemented with other required specifications.

-------Answer-------

In the beginning, the Product Backlog only lays out the initially known and best-understood requirements. The Product Backlog is a living artifact that evolves and constantly changes to identify what the product needs to be appropriate, competitive, and useful. Correct answer is 'a.'

10. The primary objective behind why a Scrum Master ensures that the Scrum Team and those interacting with the Team from the outside adhere to the Scrum rules is

a) To preserve the hierarchy of reporting and communication protocol.

b) To maximize the value created by the Scrum Team.

c) To avoid the likelihood of future audits finding any violations in implementing Scrum.

-------Answer-------

It is the Scrum Master's responsibility to ensure that outside interactions do not distract the Team from making progress in the Sprint. If beneficial, the Scrum Master can moderate these interactions to maximize the value created by the Scrum Team. Correct answer is 'b.'

#

Chapter 4 - How to execute Scrum?

The first Sprint starts after staffing the required Team roles and creating just enough Product Backlog.

Chapter 4.1 - Heart of the Execution - Sprint

What is achieved in a Sprint?

Sprint is the heart of Scrum. Each Sprint may be considered as a project with no more than a one-month horizon. The Sprint itself is an event that is the container of all the other events (Sprint Planning, Daily Scrums, Sprint Review, and Sprint Retrospective) to achieve the Product Goal. Like projects, Sprints are used to accomplish something, creating a valuable and useful product Increment, thereby progressing towards the Product Goal.

Each Sprint includes:

- A Sprint Backlog – a definition of what (is to be built), why (Sprint Goal), how (a plan that will guide building it), and optionally one or more team improvement items.
- The development work.
- The resultant product (Increment).

<u>Every Sprint always needs to produce a Done, valuable, and useable product Increment.</u>

-------------------Question- 27-a-----------------------

Sprint Backlog MUST contain at least one improvement item (identified from the Sprint Retrospective)

a) True.

b) False.

-------Answer-------

In Sprint Retrospective, the Scrum Team identifies the most helpful changes (in is way of working) to improve its effectiveness. The most impactful improvements are addressed as soon as possible. They may even be added to the Sprint Backlog for the next Sprint, but it is not mandatory. Correct answer is 'b.'

-------------------Question- 27-a-----------------------

What are the boundaries of a Sprint?

Each Sprint begins right after the conclusion of the previous Sprint. Each Sprint ends with its Retrospective, which is the last event.

How is the duration or length of the Sprint finalized?

The Scrum Team chooses a maximum duration of one calendar month or less. When a Sprint's horizon is too long, the definition of what is being built may change, complexity may rise, and risk may increase. Sprints enable predictability by ensuring inspection and adaptation of progress towards a Sprint Goal at least every calendar month. Sprints also limit risks to one calendar month of cost.

The Sprint duration is decided by the Scrum Team after considering

- The need of the Product Owner to limit business risks,
- The need of the Developers to synchronize the development work with other business events, and
- The time needed for the Developers to meet all conditions required by the Definition of Done.

Can the Sprint duration be changed later?

Sprints are best when they have consistent durations throughout a development effort, unless there is a good reason to modify. The Scrum Team should inspect and modify the duration of the Sprint only in the last Sprint event, the Retrospective, and not in the middle of a Sprint.

The Sprint is active as long as the Sprint Goal is valid. The Product Owner is the authority to decide if the goal is obsolete or not.

What work is performed within a Sprint?

- Within the Sprint, the Developers spend most of the time on development work: The team works to complete the selected Product Backlog Items to reach the Sprint Goal. Traditionally, the development work is decomposed into work tasks, and a Project Manager assigns these work tasks to the next team leader or a team member. In Scrum, the Developers organizes and manages its own work. They decide their work plan and collaboratively assign the work among themselves.

- Within the Sprint, the Developers spend less than 10% of their time on Product Backlog Refinement: The Product Owner does not interfere in the Developers' work. However, they engage the Developers in refining the Product Backlog Items. This is the only activity in the Sprint where the Developers perform an activity that is not within the current scope of the Sprint Goal. It is important to continuously refine the Product Backlog Items so there will always be some "Ready" items that can be taken up in the next Sprint.

-------------------Question- 27-b-----------------------

In the first few Sprints, the Developers are expected to focus on

a) Setting up the basic infrastructure needed for subsequent Sprint work.

b) Reviewing and baselining the project plan so the changes can be controlled.

c) Iteratively refining the requirements and obtaining sign-off from the Product Owner.

d) Delivering a valuable and useful product Increment.

e) All of the above.

-------Answer-------

The Developers must try to deliver at least one piece of functionality NOT only in the first few Sprints but in every single Sprint. Correct answer is 'd.'

-------------------Question- 27-b-----------------------

A pictorial representation of a sample Sprint with a one-week duration is shown in Figure 13.

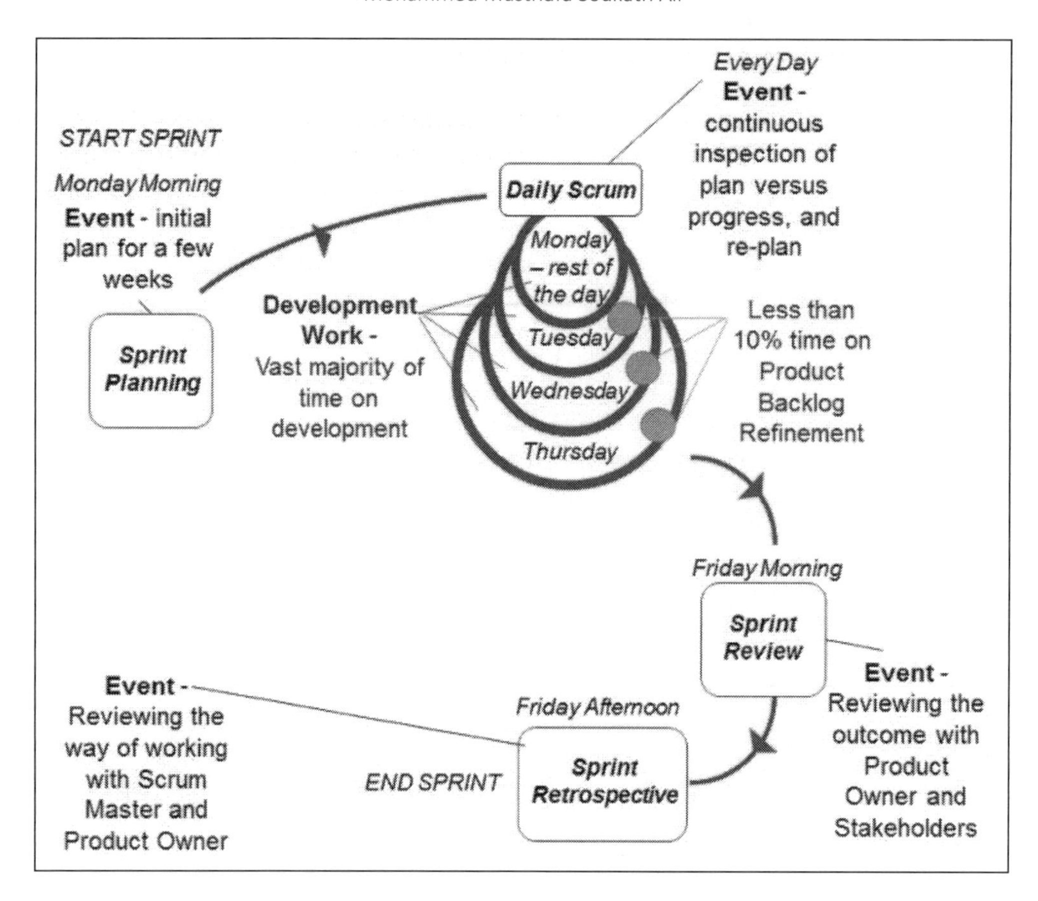

Fig. 13- A Sprint of one-week duration

Do the Developers have any flexibility to change the scope of a Sprint?

The Sprint Goal provides some flexibility to the Developers. The scope may be clarified and re-negotiated between the Product Owner and Developers as more is learned. Accordingly, the Product Backlog Items for the current Sprint can be modified as long as these changes do not endanger the Sprint Goal. If the Sprint Goal is impacted due to the changes, the Sprint will be cancelled.

Do the Developers have any flexibility to change the Definition of Done of a Sprint?

Once the Sprint is in motion, the Definition of Done for the current Sprint cannot be changed if the change will lead to a decrease in product quality. The team can inspect the need for changing the Definition of Done during the last event of the Sprint, the Retrospective, and adjust as needed for the subsequent Sprints.

Do the Developers have any flexibility to change the duration of a Sprint?

The Sprint duration is constant and is finished on a preset end date. Once the Sprint is in motion, the end date of the current Sprint is not changed. The team can inspect the need for changing the Sprint duration in the Retrospective.

The only exception is - a Sprint can be cancelled before the Sprint time-box is over. A Sprint would be cancelled if the Sprint Goal becomes obsolete. This might occur if the company changes direction or if market or technology conditions change. However, only the Product Owner has the authority to cancel the Sprint, although they may do so under the influence of the stakeholders, Developers, or Scrum Master. In general, a Sprint should be cancelled if it no longer makes sense given the circumstances. But, due to the short duration of Sprints, cancellation rarely makes sense.

Sprint cancellations consume resources since everyone has to regroup in another Sprint Planning to start a fresh Sprint. Sprint cancellations are often traumatic to the Scrum Team and are uncommon.

What happens to the work in progress when the Sprint is cancelled?

When a Sprint is cancelled, any completed and Done Product Backlog Items are reviewed. If part of the work is valuable and useful, the Product Owner typically accepts it. All incomplete Product Backlog Items are re-estimated and put back into the Product Backlog. The work done on them depreciates quickly and must be frequently re-estimated.

Summary

- The heart of Scrum is a Sprint, during which a Done, valuable, and useable product Increment is created. This allows the Scrum Teams to inspect and adapt of progress toward a Product Goal at least every calendar month (maximum Sprint length).

- Each Sprint may be considered a project with no more than a one-month horizon. When a Sprint's horizon is too long the definition of what is being built may change, complexity may rise, and risk may increase.

- Sprints are best when they have consistent durations throughout a development effort.

- Each Sprint has a definition of what (is to be built), why (Sprint Goal), how (a design and flexible plan that will guide building it), the work itself, and the resultant product.

- During the Sprint, the scope may be clarified and re-negotiated between the Product Owner and Developers as more is learned.

- No changes are made that would endanger the Sprint Goal, and the quality goals do not decrease.

- A Sprint would be cancelled if the Sprint Goal becomes obsolete.

- Only the Product Owner has the authority to cancel the Sprint, although they may do so under the influence of others.

- When a Sprint is cancelled, the Product Owner typically accepts the completed work that is valuable and useful. All incomplete Product Backlog Items are re-estimated and put back into the Product Backlog.

#

Chapter 4.2 - Sprint starts with - Sprint Planning

What is achieved in Sprint Planning?

The team selects Product Backlog Items to forecast the work for the Sprint. Sprint Planning creates a definition of what is to be built and a flexible plan. The outcome is a Sprint Backlog that contains Product Backlog Items, a plan for delivering them, and one or more team improvement items. In addition to the Sprint Backlog, the team also creates a Sprint Goal.

This plan is created by the collaborative work of the entire Scrum Team.

The set of selected Product Backlog Items that the Developers think they can complete, and the associated work is called a **forecast** of functionality. The term forecast was called a **commitment** in older versions of The Scrum Guide.

Who leads and controls Sprint Planning?

The entire Scrum Team gathers for the Sprint Planning. If the Scrum Team invited any external technical/domain experts, they participate as well. No individual leads or controls this meeting.

The Scrum Master ensures that the event takes place and that attendants understand its purpose. The planning should answer three questions:

- Topic One: Why is this Sprint valuable? (Sprint Goal)
- Topic Two: What can be Done this Sprint?
- Topic Three: How will the chosen work get done?

The Scrum Master teaches the Scrum Team to keep it within the time-box. Its duration is up to eight hours for a one-month Sprint. For shorter Sprints, it is usually shorter.

What happens in Sprint Planning?

Under the first topic of Sprint Planning, the Scrum Team crafts a Sprint Goal, that communicates why the Sprint is valuable to stakeholders. The Sprint Goal is an objective that will be met within the Sprint through the implementation of the Sprint Backlog. The Sprint Goal provides guidance to the Developers on why it is building the Increment.

The Sprint Goal is a tool for team coherence. The selected Product Backlog Items deliver one coherent function, which can be the Sprint Goal. The coherence between the Product Backlog Items is made transparent by the Sprint Goal. This is important because it provides an opportunity for team members to work together and offers some flexibility of adjusting the Product Backlog Items when required. Lack of coherence may lead the team to working on separate initiatives. So, the Sprint Goal is a tool to verify and create this coherence. As the Developers work, they keep the Sprint Goal in mind. In order to satisfy the Sprint Goal, it implements the functionality and technology. If the work turns out to be different than what the Scrum Team expected, they collaborate with the Product Owner to negotiate the scope of the Sprint Backlog within the Sprint.

Under the Topic Two, the Scrum Team selects the Product Backlog Items for the current Sprint. The Product Owner discusses the objective that the Sprint should achieve and the Product Backlog Items that, if completed in the Sprint, would achieve the Sprint Goal. The Product Owner will introduce the Product Backlog Items that are "ready" - already refined to a state agreed upon by the Developers and Product Owner. Just in time refinement will lead to poorly understood product needs and result in poor Sprint Planning. The number of items selected from the Product Backlog for the Sprint is solely up to the Developers. Selecting how much can be completed within a Sprint may be challenging. However, the more the Developers know about their past performance, their upcoming capacity, and their Definition of Done, the more confident they will be in their Sprint forecasts. The selected items are moved to the Sprint Backlog.

Having selected the Product Backlog Items and the Sprint Goal, the team moves to Topic Three, the third phase of Sprint Planning. The Developers decide how it will build this functionality into a Done product Increment during the Sprint.

The Developers usually start by designing the system and then identifies the work needed to convert the selected Product Backlog Items into a working product Increment. Based on the estimate of the work, the Developers plan enough work to match the available capacity. If the Developers determine it has too much or too little work, it may renegotiate the selected Product Backlog Items with the Product Owner.

The Sprint Goal, the Product Backlog Items selected for this Sprint, and the plan for delivering them is called the Sprint Backlog.

By the end of Sprint Planning, the Developers should be able to explain to the Product Owner and Scrum Master how it intends to work as a self-managing team to accomplish the Sprint Goal and create the anticipated Increment.

-------------------Question- 28-------------------------

Which of the following does the Sprint Goal provide?

a) Guidance to the team on why it is building the Increment

b) Flexibility to the team about the functionalities implemented in this Sprint

c) Coherence so that team members can work together

d) It communicates why the Sprint is valuable to stakeholders

e) All the above

-------Answer-------

The Sprint Goal provides guidance to the Developers. It also provides some flexibility regarding the functionality implemented within the Sprint. The selected Product Backlog Items deliver one coherent function, which can be the Sprint Goal. The Sprint Goal can be any other coherence that causes the Scrum Team to work together toward a common but specific goal. Sprint Goal also communicates why the current Sprint is valuable to stakeholders. Correct answer is 'e.'

-------------------Question- 28-------------------------

Should the entire Sprint work be decomposed within Sprint Planning?

Not necessarily. The Developers decompose the work planned for the first few days of the Sprint by the end of this meeting, often to units of one day or less. Developers collaboratively assign these work units among themselves.

-------------------Question- 29-------------------------

Which estimation unit must be used by the Developers for the work needed to convert the selected Product Backlog Items into a working product Increment?

a) Function Points

b) Ideal Hours

c) Story Points

d) Any useful sizing technique

-------Answer-------

The work can be of varying size or estimated effort. Correct answer is 'd.'

-------------------Question- 29-------------------------

Note the emphasis on the rule that only the Developers finalize how much they can develop within that Sprint. No one else can persuade them with different expectations.

Summary

- In Sprint Planning, the plan for the Sprint work is created by the collaborative work of the entire Scrum Team.
- Sprint Planning is time-boxed to a maximum of eight hours for a one-month Sprint. For shorter Sprints, the event is usually shorter.
- In Topic One, the Scrum Team crafts a Sprint Goal.
- In Topic Two, the Developers work to forecast the functionality that can be done in this Sprint.
- The number of items selected from the Product Backlog for the Sprint is solely up to the Developers.
- The Sprint Goal is an objective that will be met within the Sprint through the implementation of the selected Product Backlog Items.
- In Topic Three, the Developers decides how it will build this functionality into a Done product Increment.
- The Sprint Goal, the Product Backlog Items selected for this Sprint and the plan for delivering them is called the Sprint Backlog.
- Work planned for the first days of the Sprint is decomposed by the end of this meeting, often to units of one day or less.
- The Developers undertake the work in the Sprint Backlog.
- If the work turns out to be different than the Developers expected, they collaborate with the Product Owner to negotiate the scope of the Sprint Backlog.

#

Chapter 4.3 - Planning produces a Sprint Backlog and Sprint Goal

What is the purpose of the Sprint Backlog?

The Sprint Backlog increases the transparency of information about work planned and performed during the Sprint.

It is a highly visible, real-time picture of the work that the Developers plan to accomplish during the Sprint.

What is the content and format?

Content: Sprint Goal (Why), Product Backlog Items in scope for the Sprint (What) and the Developers' plan for realizing the Sprint Goal (How). Each plan item has a description and estimate. The Sprint Backlog is also called a Forecast. Sprint Backlog also may contain improvement required in the team's way of working.

Format: Nothing specific, but each item must have a description and estimate. Figure 14 shows a sample Sprint Backlog.

Product Backlog Item	Value	Effort	Order	Sprint Work Items	Estimate (Hours)	Remaining Hours	Worked by
A registered customer can place request online to increase credit limit	50	21	Critical - 2	Update the Credit Account Database with two more fields	2	2	Open
				Create one screen with two fields to capture the request and validate it	8	8	Open
				Integrate Screen with database through existing service	6	6	Open
				Perform testing as needed by definition of done	8	8	Open
A registered customer can place request to increase credit limit from a smartphone with <xx> O/S	30	13	Important - 1	Refactor the database to capture channel information	1	1	Open
				Add one screen with two fields to capture the request and validate it	4	4	Open
				Integrate Smart phone application changes with middle layer service	4	4	Open
				Regression test online screen	1	1	Open
				Perform testing as needed by definition of done	6	6	Open

Fig. 14- Sample Sprint Backlog

What is the lifecycle of the Sprint Backlog?

The Sprint Backlog is maintained from Sprint Planning until the Sprint Review. Although it starts with just enough detail to get the Development Work started, the Sprint Backlog emerges during the Sprint as the Developers learn more about the work and modifies the Sprint Backlog accordingly.

The Sprint Backlog belongs solely to the Developers. Only the Developers can change the Sprint Backlog.

As work is performed or completed, the Developers update the estimated remaining work.

What happens if the team finds surprises about the work during the Sprint?

The Developers work through the plan and learns more about the work needed to achieve the Sprint Goal. The team may find that they under- or overestimated a Product Backlog Item. It is absolutely normal. As new work is identified, the Developers add it to the Sprint Backlog. When elements of the plan are deemed unnecessary, they are removed. The Developers modify the Sprint Backlog throughout the Sprint, and the Sprint Backlog emerges during the Sprint.

-------------------Question- 30--------------------------

The Sprint Backlog is modified throughout the Sprint. As soon as a new task is identified,

a) The Product Owner adds it to the Sprint Backlog and communicates it to the Scrum Team.

b) The Scrum Master adds it to the Sprint Backlog and communicates it to the Scrum Team.

c) The Developers add it to the Sprint Backlog and communicate it to the Scrum Team.

-------Answer-------

The Sprint Backlog belongs solely to the Developers. Correct answer is 'c.'

-------------------Question- 30--------------------------

Rules

- The Developers only can finalize what is part of the Sprint Backlog.
- Each plan item in the Sprint Backlog is preferably decomposed to less than a day's work.
- The Sprint Backlog is collectively owned by the Developers. There is no sole ownership of a Sprint Backlog Item by an individual team member, although they may work individually on an item. This is to ensure that there is increased transparency of work within the team, without an individual boundary.
- The estimated total remaining time is calculated at least once a day before the Daily Scrum, but usually more often than that.

-------------------Question- 31--------------------------

A Scrum Team decides to divide the Sprint Backlog and assign ownership of every Sprint Backlog Item to separate individuals on the team. The Scrum Master

a) Should encourage this practice as it increases productivity.

b) Should coach the team to collectively take ownership of the Sprint Backlog Items even though an individual works on a specific item.

c) Should encourage this practice as it increases individual accountability.

-------Answer-------

The Sprint Backlog is collectively owned by the Developers. Correct answer is 'b.'

-------------------Question- 31--------------------------

Planning produces the Sprint Goal

The Sprint Goal is not maintained as a separate artifact in Scrum. It is a 'commitment' attached to the Sprint Backlog. Considering the importance and significance of the Sprint Goal, the author has pulled it out to highlight its special importance.

It is usually made available as publicly radiated information for the Scrum Team.

The Sprint Goal is part of the Sprint Backlog. While the Sprint Goal is not changed, the Sprint Backlog items can be negotiated anytime during the Sprint.

When the Sprint Goal becomes obsolete, the Sprint is automatically cancelled. There are various reasons for the Sprint Goal to become invalid, such as market changes, organizational changes, etc.

-------------------Question- 32-------------------------

During the Sprint, while the Sprint Backlog can be modified as more is learned, no changes are made that would endanger the Sprint Goal.

a) True

b) False

-------Answer-------

If the Sprint Goal is endangered, it would lead to cancellation of the Sprint. Correct answer is 'a.'

-------------------Question- 32-------------------------

Summary

- The Sprint Backlog is a forecast by the Developers about what functionality will be developed in the Sprint, the Sprint Goal, and the work needed to deliver that functionality, and one or more team improvement items.
- The Sprint Backlog is a plan with enough detail so that progress against this plan can be understood in the Daily Scrum.
- The Developers modify the Sprint Backlog throughout the Sprint, and the Sprint Backlog emerges during the Sprint.
- As new work is identified, the Developers add it to the Sprint Backlog. As work is performed or completed, the estimated remaining work is updated.
- The Sprint Goal communicates why the current Sprint is valuable to stakeholders. The Sprint Goal can be one coherent business function or any other coherence that causes the Developers to work together rather than on separate initiatives.
- The Sprint Goal also provides some flexibility to the Developers regarding the Product Backlog Items.
- Only the Developers can change the Sprint Backlog.

#

Chapter 4.4 - Development Work

--------------------------DE-TOUR--------------------------

In this chapter, the following are included for clarity and context. They are not part of The Scrum Guide:
New team's ability to produce an Increment in short Sprints, Continuous planning by the Scrum Team

--------------------------DE-TOUR--------------------------

At the end of Sprint Planning, the Developers start the development work. The team discusses the work items in the Sprint Backlog and collaboratively decides who will work on what task. Though they may work individually on an item, there is no sole ownership of a Sprint Backlog Item by an individual team member. This is to ensure that there is increased transparency of work within the team, without any individual boundary.

Development work may include the necessary product engineering practices. The Developers are expected to be cross-functional enough to have all the skills needed for engineering the product Increment without any external help.

A Product Backlog Item is considered to be completely done only if it meets the conditions defined in the Definition of Done.

The Sprint is too short. How can a Scrum Team produce a fully functional Increment?

This is one of the major problems new Scrum Teams face. Usually there are many constraints to creating a fully functional Increment due to the way an Organization may work today, and many of these constraints will block the Scrum Team's progress. Some examples in the software development process include:

- The development cycle needs to be followed by a lengthy testing cycle.
- The required skills are not available within the team.
- The Product Backlog Item was not sufficiently decomposed.

Following the Scrum rules helps to avoid many such blockers. For example, staffing the team such that it becomes cross-functional eliminates the problem of lack of required skills. Refining the Product Backlog ahead of Sprint Planning helps to have granular Product Backlog Items that are less complex to develop.

Yet the teams will face many issues that cannot be readily addressed by just following Scrum. The lengthy test cycle is an example.

One of the primary responsibilities of the Scrum Master for new teams is to coach them in the art of working to produce a useable Product Increment within short Sprints. In addition, the Scrum Master needs to coach the team to identify issues every day and learn to resolve those issues by themselves. The road to excellence is a journey, and the team needs to go through this initial learning with the Scrum Master's help.

When there are issues that are outside the team's influence, they are called impediments. The Scrum Master needs to own these impediments and resolve them. For example, the lengthy testing cycle example above may be an organizational-level constraint that the Scrum Master needs to work with the larger Organization to find an alternative solution.

Self-managing teams – How would they synchronize and track the collective progress without a manager?

Though the Project Manager role is not there, the traditional job of project planning and control is still performed. The difference is that the planning is continuous, and the entire team performs the planning and control instead of leaving that to a management position.

The Developers

- Plan how the development work will be performed in terms of decomposing the work into work units. Decomposition for the first days of the Sprint happens in Sprint Planning.

- Track the progress against the Sprint plan. The Developers plan work for the next 24 hours in the Daily Scrum, by inspecting the work since the last Daily Scrum and forecasting upcoming Sprint work. The team may use tools like a Sprint Burn Down that shows how much work is left against the remaining time.

- Take control and communicates actions if there is a variance between actual progress versus the plan.

- Continue this tracking->re-planning->communicating->controlling cycle every day.

-------------------Question- 33--------------------------

Since the team is self-managing in Scrum and they manage their own work, they do not need any planning to perform their work.

a) True

b) False

-------Answer-------

There are still planning activities in Scrum. Though there is no detailed long-term planning for a complete project, the planning is done for short-term Sprints. Within Sprint, the Developers further plans work for the next 24 hours in the Daily Scrum, by inspecting the work since the last Daily Scrum and forecasting upcoming Sprint work.

This is a more realistic approach, and continuous planning allows the team to be agile enough to pro-actively look for changes and respond to the changes. Correct answer is 'b.'

-------------------Question- 33--------------------------

Summary

This chapter is too short to review. Reread, if necessary.

#

Chapter 4.5 – Developers review progress every day in Daily Scrum

--------------------------DE-TOUR--------------------------

In this chapter, the following are included for clarity and context. They are not part of The Scrum Guide:
Sprint Burn-down Chart, Guidance on Issues and Impediments, Backlog of Impediments

--------------------------DE-TOUR--------------------------

What is achieved in the Daily Scrum?

In the Daily Scrum, the Developers synchronize the on-going activities and create a plan for the next 24 hours to drive its development work. Also, any impediments are updated to the Backlog of impediments and made transparent so others including the Scrum Master will know the details even if they do not attend the Daily Scrum.

Who sets up the meeting?

As a self-managing team, the Developers are responsible for conducting the Daily Scrum. The Daily Scrum is held at the same time and place each day to reduce complexity - doing something regularly at the same time and place creates an instinctive habit for the team. Moreover, it simplifies the logistics. The Daily Scrum is a 15-minute time-boxed event irrespective of the size of the team or duration of the Sprint. Once initially agreed upon, the Developers gather for the meeting at the same time and place every day without any special arrangements typically needed for traditional meetings.

Who leads and controls the Daily Scrum?

The Daily Scrum is an internal meeting for the Developers. If others are present, the Scrum Master ensures that they do not disrupt the meeting. The Scrum Master or any individual does not lead or control the event. The Scrum Master

- ensures that the Developers has the meeting and
- teaches the Developers to keep the Daily Scrum within the 15-minute time-box.

What happens if one or more Developers are out of the office?

The Daily Scrum goes on irrespective of minor developments such as someone being away. In fact, the Daily Scrum is a key inspect and adapt opportunity used to identify any issues like an unexpected absence and adapt the next 24-hour plan to respond. This daily sync up increases the possibility of acting on issues sooner and hence optimizes the probability of getting the work done to meet the Sprint Goal.

Why is the Daily Scrum a key inspect and adapt meeting?

Every day the Developers should understand how they intend to work together as a self-managing team to accomplish the Sprint Goal and create the anticipated Increment by the end of the Sprint.

Daily Scrums improve communication, eliminate other meetings, identify impediments to development for removal, highlight and promote quick decision-making, and improve the Developers' level of knowledge.

-------------------Question- 34-a-----------------------

Not having the Daily Scrum will

a) Reduce the opportunity to create a status report.

b) Reduce the speed of development work.

c) Reduce the transparency of overall progress.

-------Answer-------

Failure to include any of the Scrum events (including the Daily Scrum) results in reduced transparency and is a lost opportunity to inspect and adapt. Correct answer is 'c.'

------------------Question- 34-a-----------------------

What happens in the Daily Scrum?

There is no standard format for the Daily Scrum. The structure of the meeting is set by the Developers and can be conducted in different ways if it focuses on progress toward the Sprint Goal. Here is an example of how a team starts the event:

The team starts by discussing the progress thus far. This is done by inspecting the work completed since the last Daily Scrum. Since the Sprint Backlog shows the work units remaining for each item, the team can understand the progress made in the previous 24 hours.

To synchronize between each other, some Developers will use questions, some will be more discussion based. Here is an example of what might be used:

- *What did I do yesterday that helped the Scrum Team meet the Sprint Goal?*
- *What will I do today to help the Scrum Team meet the Sprint Goal?*
- *Do I see any impediment that prevents me or the Scrum Team from meeting the Sprint Goal?*

------------------Question- 34-b------------------------

Developers must use three standard questions (yesterday, today, impediment) in Daily Scrum.

a) True.

b) False.

-------Answer-------

In earlier versions of The Scrum Guide, it was mandatory and now it is only a suggestion. Correct answer is 'b.'

------------------Question- 34-b------------------------

This is not a status meeting run by a manager, so the team does not resort to any particular style such as one individual asking these questions of the others.

The team agrees on some ordering, preferably based on the Sprint Backlog plan, and voluntarily provides the status to rest of the team. Some teams may pass an object like a ball to each other, indicating that whoever has the ball talks. Such tactics are not part of Scrum but may be used by the team if they find them useful to increase collaboration and self-management. If there are issues that the team can mutually solve with each other's help, the team agrees to collaborate on them. If there are impediments outside of their influence blocking progress, they are captured and made transparent to the Scrum Master.

------------------Question- 35------------------------

During the Daily Scrum, a team member says he does not know when his task will be complete.

a) It is acceptable as the Sprint Review date is far away.

b) Replace the team member with a new team member.

c) The Developers should collaborate to plan alternative steps such as pairing the member with someone else to eliminate the risk of not meeting the Sprint Goal.

d) The Scrum Master should mentor the team member on how to estimate the task.

-------Answer-------

As a self-managing team, it is the team's collective responsibility to immediately take steps to resolve issues and meet the Sprint Goal. They still have to discuss how to improve the team member's abilities, but it is a topic for discussion later in the Sprint Retrospective. This issue is well within the influence of the Developers to solve and hence the Scrum Master's help is not needed. Correct answer is 'c.'

-------------------Question- 35--------------------------

How do the Developers monitor the Sprint progress?

The Sprint Backlog is a plan with enough detail to be a reference. Using this reference, any changes in progress can be understood on a regular basis. This change in progress is inspected and any deviations are acted upon.

The team may optionally use a technique like a Sprint Burn-down to project the trend of completion. A Sprint Burn-down is not mandatory but may be used if the team finds value.

A Sprint Burn-down Chart is a tool that shows the remaining work in the Sprint. Its format is usually a graph containing the days on the x-axis and the work on the y-axis. Using a Sprint Burn-down Chart, the team tracks the estimated remaining work to meet the Sprint Goal. Based on the findings, the team forecasts the work that can be done before the next Daily Scrum. Any adjustments identified for the next 24 hours are updated in the Sprint Backlog.

Figure 15 shows a burn-down chart. This is how it appears for a fictitious Scrum Team, when they look at it at the end of 6th day of their Sprint.

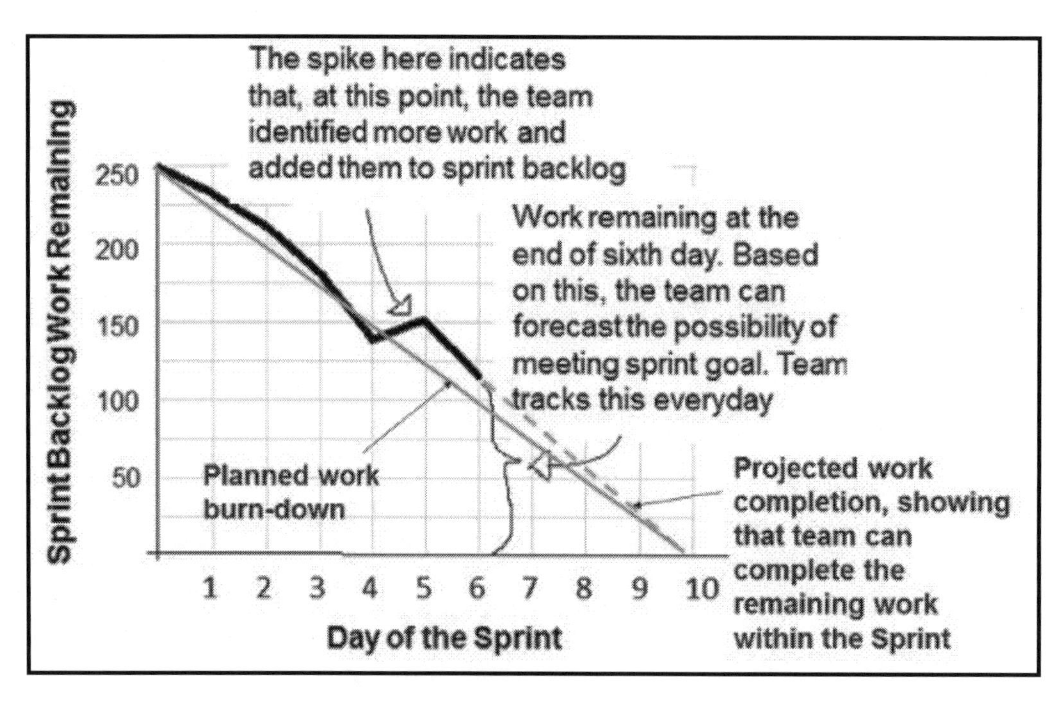

Fig. 15- Sample of a Sprint Burn-down Chart

-------------------Question- 36-------------------------

In Scrum, the usage of a forecast tool like a Burn-down Chart is a foolproof way of estimating the completion of product development.

a) Incorrect. The Burn-up Chart is the better alternative.

b) Incorrect. Such practices, though useful to some extent, do not replace the importance of empiricism.

c) Correct. If the Scrum Team is highly disciplined in updating the Burn-down Chart, then this can be true.

-------Answer-------

Burn-up and Burn-down Charts have proven useful. However, these do not replace the importance of empiricism. In complex environments, what will happen is unknown. Only what has happened may be used for forward-looking decision-making. Correct answer is 'b.'

-------------------Question- 36-------------------------

Is the Progress calculated only during the Daily Scrum?

Not necessarily. At any point during a Sprint, the total work remaining in the Sprint Backlog can be summed. At a minimum, the Developers track this total work remaining in every Daily Scrum to project the likelihood of achieving the Sprint Goal. The Developers can track the remaining work at any time throughout the Sprint.

How do the Developers handle issues and impediments against the Sprint Goal?

The highest priority of the Developers is to complete the Sprint Goal. An impediment refers to any problem faced by the Developers that stops or blocks their planned progress towards the Sprint Goal. Impediments threaten the completion of the Sprint by the pre-set date.

In Scrum, impediments are continuously identified throughout the Sprint, and they are made transparent during the Daily Scrum. The following are guidelines that the Developers follow whenever they face impediments:

1. Make impediments visible: As soon as someone identifies an impediment, they make it transparent to other team members through some communication or at the Daily Scrum, whichever is earlier.

2. Solve the impediment if it is within their influence: If there are impediments that are within the influence of the team to resolve, they need to be resolved by the team either directly or using workarounds.

Note: The Developers can reach out to technical/domain experts outside the Scrum Team for help. However, later in the Sprint Retrospective, they need to inspect why they could not do it themselves. They need to identify the improvements necessary to make them truly cross-functional and self-sufficient.

3. Use the Scrum Master's help for impediments outside the team's influence: If the impediments need broader collaboration outside the team, the team should seek the Scrum Master's help.

4. If the impediments still persist, involve the Product Owner: After the above steps, at the earliest indication that the team or Scrum Master cannot make progress on the impediments, the team should involve the Product Owner to discuss the alternatives.

If the Scrum Master does not attend the Daily Sprint, how would they know about impediments?

The Developers must participate in the Daily Scrum. However, the Scrum Master can choose to attend to ensure Scrum is correctly understood and enacted. They can also facilitate the Daily Scrum at the team's request.

The question is – If the Scrum Master does not attend the Daily Sprint, how would they know about impediments? This question assumes that impediments are brought out only during the Daily Scrum. That is incorrect. A team needs to bring out any impediment as soon as they know they need help.

The team will transparently make the impediments visible for anyone to see and also approach the Scrum Master when they know they need help.

What if the team identifies adjustments which are needed for the rest of the Sprint plan and not just the next 24 hours?

If the Daily Scrum exposes the need to re-plan the rest of the Sprint, not just 24 hours, the Developers often meet immediately after the Daily Scrum for detailed discussions or to adapt or re-plan the rest of the Sprint's work. Regardless, the Developers always close the Daily Scrum within 15 minutes.

At the end of each Daily Scrum, the Developers will have a plan for the next 24 hours updated in the Sprint Backlog. This plan will contain the most important work that needs to be done in the next 24 hours. Also, the Backlog of impediments is updated and made transparent so others including the Scrum Master will know even if they do not attend the Daily Scrum.

By performing the development work, the Developers reach the Sprint Goal by producing a product Increment.

Summary

- The Daily Scrum is a 15-minute time-boxed event for the Developers to synchronize activities and create a plan for the next 24 hours. This optimizes team collaboration and performance by inspecting the work since the last Daily Scrum and forecasting upcoming Sprint work.

- The Scrum Master ensures that the Developers have the meeting, but the Developers are responsible for conducting the Daily Scrum. The Scrum Master teaches the Developers to keep the Daily Scrum within the 15-minute time-box.

- The Developers use the Daily Scrum to inspect progress toward the Sprint Goal and to inspect how progress is trending toward completing the work in the Sprint Backlog. The Daily Scrum optimizes the probability that the Developers will meet the Sprint Goal.

- The Daily Scrum is held at the same time and place each day to reduce complexity.

- The Daily Scrum is an internal meeting for the Developers. If others are present, the Scrum Master ensures that they do not disrupt the meeting.

- There is no standard format for the Daily Scrum. The structure of the meeting is set by the Developers and can be conducted in different ways if it focuses on progress toward the Sprint Goal. Some Scrum Teams will use questions, some will be more discussion based.

- Every day, the Developers should understand how they plan to work together as a self-managing team to accomplish the Sprint Goal and create the anticipated Increment by the end of the Sprint.

- The Developers inspect the progress towards the Sprint Goal and the trend towards completing the Sprint Backlog.

- The Developers (not everyone but as required) often meet immediately after the Daily Scrum for detailed discussions, or to adapt, or replan, the rest of the Sprint's work.

- This is a key inspect and adapt meeting. Daily Scrums improve communication, eliminate other meetings, identify impediments to development for removal, highlight and promote quick decision-making, and improve the Developers' level of knowledge.

#

Chapter 4.6 – Team produces the Increment

What is the significance of the Increment?

The Increment is the most important artifact of all because it is the mark of the progress towards a business goal or vision. This is the end result of work completed. The Increment is not just the outcome of the latest Sprint. It is the sum of all the Product Backlog Items completed during a Sprint and the value of the Increments of all previous Sprints.

Format: Note that this artifact is not a document but a working product. The new Increment must be Done, which means it must be in a useable condition and meet the Scrum Team's Definition of Done. It must be in a useable condition regardless of whether the Product Owner decides to actually release it.

The team has a Product Backlog Item that is almost done. Can it be demonstrated in the Sprint Review?

No. Any Product Backlog Item is incomplete if it does not meet the Definition of Done. Delivering incomplete items to the Sprint Review reduces the transparency of real progress and increases risks due to unknown work that is yet to be completed.

-------------------Question- 37-a-------------------------

Within every Sprint, the working Increment should be tested progressively starting from unit testing, then integration testing, and finally user acceptance testing.

a) Yes. It is the prescribed method.

b) No. The test strategy is decided by the Quality Assurance Lead in the team.

c) Not necessarily. It is up to the team to find the best approach to testing.

d) Incorrect. It should also include non-functional testing.

-------Answer-------

While the team needs to ensure that each Increment is thoroughly tested, all Increments work together and meet the Definition of Done, it is up to the team to find the best approach to achieve this. Correct answer is 'c.'

-------------------Question- 37-a-------------------------

-------------------Question- 37-b-------------------------

The increment is a step towards

a) A vision or goal.

b) Sprint completion or project closure.

c) Sign-off or phase completion.

-------Answer-------

The Increment is the sum of all the Product Backlog items completed during a Sprint and the value of the increments of all previous Sprints. An increment is a body of inspectable, done work that supports empiricism at the end of the Sprint. The increment is a step toward a vision or goal. Correct answer is 'a.'

-------------------Question- 37-b-------------------------

The Increment is delivered to the Product Owner and the stakeholders for their inspection in the Sprint Review. The Sprint Review is specifically kept informal to foster the participants to bring in the highest amount of transparency to what was achieved and openly discuss how to best leverage this new understanding and adapt for the next Sprint.

-------------------Question- 38-------------------------

In the Sprint Review, the presentation of the product Increment to stakeholders is

a) To get the Sprint completion sign-off.

b) To provide the status of the project.

c) To elicit feedback.

-------Answer-------

The Sprint Review is an informal meeting, not a status meeting, and the presentation of the Increment is intended to elicit feedback and foster collaboration. Correct answer is 'c.'

-------------------Question- 38-------------------------

Summary

This chapter is too short to review. Reread, if necessary.

#

Chapter 4.7 – Stakeholders collaborate in the Sprint Review

--------------------------DE-TOUR--------------------------

In this chapter, the following are included for clarity and context. They are not part of The Scrum Guide:
Product Burn-down Chart

--------------------------DE-TOUR--------------------------

What is achieved in the Sprint Review?

The Scrum Team and the stakeholders inspect the Increment, collaborate to identify the next valuable things to do in view of the insights gained from inspection, and adapt the Product Backlog if needed. The output from this Sprint Review is valuable input to the next Sprint Planning.

Who leads and controls the Sprint Review?

The Sprint Review attendees include the Scrum Team and key stakeholders invited by the Product Owner. It is an informal meeting so that the participants can bring in the highest amount of transparency and collaboration. No individual leads or controls the meeting. The Scrum Master ensures that the event takes place and that the attendees understand its purpose and teaches all to keep it within the time-box. This is a four-hour time-boxed meeting for one-month Sprints. For shorter Sprints, the event is usually shorter.

What happens in the Sprint Review?

The Product Owner explains which Product Backlog Items are Done and which are not.

The Product Owner discusses the Product Backlog as it stands and projects the likely completion date based on progress to date.

At any point, the total work remaining to reach the goal can be summed. The Product Owner tracks this total work remaining in at least every Sprint Review. The Product Owner compares this amount with the work remaining at previous Sprint Reviews to assess progress towards completing the projected work by the desired goal time.

How does the Product Owner monitor the progress towards the goal?

Various projective practices, based on trending, have been used to forecast progress, like Burn-down Charts, Burn-up Charts, or Cumulative Flow Diagrams. The Product Owner may use them to track progress and forecast completion. These are not artifacts mandated by Scrum, but the Scrum Team may use them if the tools add value.

A Product Burn-down Chart shows the estimated total amount of work (Product Backlog Items) remaining and its evolution over time. A Product Burn-up Chart shows the amount of Product Backlog Items cumulatively Done so far and its evolution over time.

These are usually a graph containing the Sprint numbers in the x-axis and the Product Backlog work on the y-axis. These are maintained throughout the product development cycle and updated by the Product Owner during the Sprint Review.

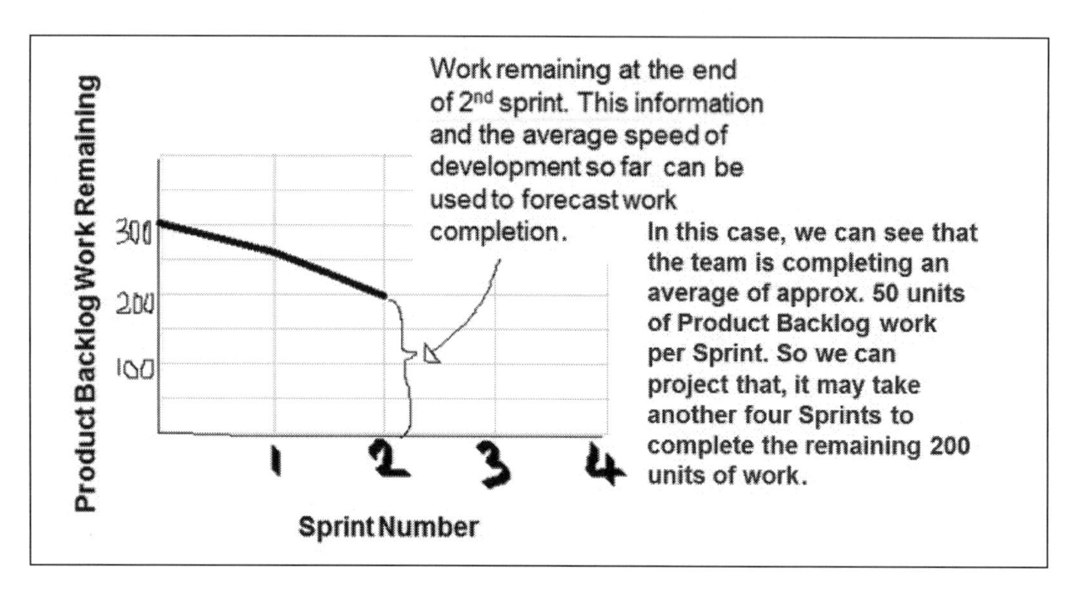

Fig. 16– Sample of a Product Burn-down Chart

Using these tools, the Product Owner compares the estimated remaining work at the Sprint Review with the same metric from the previous Sprint Review, understands the progress made towards completing the projected work by the desired time for the goal, and makes it transparent to the stakeholders.

------------------Question- 39-------------------------

The difference between the Product Burn-down Chart and the Sprint Burn-down Chart is

a) In the Product Burn-down Chart, the number of Product Backlog Items is shown on the y-axis. In the Sprint Burn-down Chart, the number of tasks is shown on the y-axis.

b) In the Product Burn-down Chart, the Sprints are shown on the x-axis. In the Sprint Burn-down Chart, the days of the Sprint are shown on the x-axis.

c) In the Product Burn-down Chart, the Sprints are shown on the y-axis. In the Sprint Burn-down Chart, the days are shown on the y-axis.

-------Answer-------

For both the Product Burn-down Chart and the Sprint Burn-down Chart, the y-axis always represents the work remaining. Note that the "number of work items (tasks)" is different from the "amount of work." Correct answer is 'b.'

------------------Question- 39-------------------------

What do the Developers present during the Sprint Review?

The Developers present what went well during the Sprint, what problems it experienced, and how those problems were resolved. These discussions are not about technical problems, but those relating to understanding the Product Backlog Items, business and product issues that were uncovered, coherence or lack of coherence between the items, conflicts with previous Sprint implementations, and so on. These discussions will add more transparency.

The Developers use most of the Sprint Review time to demonstrate the work that is Done and answers questions about the Increment. They do not demonstrate any items that are incomplete.

What is adapted in the Sprint Review?

The participants collaborate on what to do next. They review how the marketplace or potential use of the product might have changed and what are the most valuable things to do next. They also review the timeline, budget, potential capabilities, and marketplace for the next anticipated product release.

At the end, they come up with valuable input to the subsequent Sprint Planning by adapting the Product Backlog. A revised Product Backlog defines the probable Product Backlog Items for the next Sprint. However, the scope of the next Sprint is finalized only in the next Sprint Planning. The Product Backlog may also be adjusted overall to meet new opportunities.

-------------------Question- 40-------------------------

Select all that apply. The Sprint Review is an event that requires

a) The Product Owner's sign-off.
b) Active participation of the stakholders invited by the Product Owner.
c) Transition sign-off.
d) Inspection and adaptation activities.

-------Answer-------

Sprint Review doesn't require formal sign-offs as it is an informal meeting to elicit feedback and foster active collaboration between all the participants. Correct answers are 'b' and 'd.'

-------------------Question- 40-------------------------

Summary

- A Sprint Review is held at the end of the Sprint to inspect the Increment. The Product Backlog is adapted if needed.
- This is a four-hour time-boxed meeting for one-month Sprints. For shorter Sprints, the event is usually shorter.
- This is an informal meeting to elicit feedback and foster collaboration.
- The Scrum Team and key stakeholders participate and collaborate.
- The Developers demonstrate the work that is Done.
- The Product Owner projects the likely completion dates.
- The entire group collaborates on what to do next, a valuable input to the subsequent Sprint Planning.
- They review the timeline, budget, and potential capabilities and marketplace for the next Increment.
- The result of the Sprint Review is a revised/adjusted Product Backlog.

#

Chapter 4.8 – Sprint ends with the Sprint Retrospective

What is achieved in the Sprint Retrospective?

The Scrum Team is a self-managing team that continuously improves its way of working Sprint to Sprint without any external push. Although improvements may be implemented at any time, the Sprint Retrospective provides a formal opportunity to focus on inspecting the Scrum Team itself.

-------------------Question- 41--------------------------

What is inspected in the Sprint Retrospective?

a) Sprint Improvement Plan
b) Scrum Team
c) Sprint Backlog

-------Answer-------

The Scrum Team identifies the most helpful changes to improve its own effectiveness. Correct answer is 'b.'

-------------------Question- 41--------------------------

Who leads and controls the Sprint Retrospective?

The Sprint Retrospective is an opportunity for the Scrum Team to inspect itself and create a plan for improvement to be enacted during the next Sprint.

The Scrum Master participates as a peer team member in the meeting from the accountability of the Scrum process. The Scrum Master or any individual does not lead or control the event. The Scrum Master ensures that the event takes place and that attendants understand its purpose and teaches all to keep it within the time-box. This is a three-hour time-boxed meeting for one-month Sprints. For shorter Sprints, the event is usually shorter.

What happens in the Sprint Retrospective?

The Scrum Team inspects how the last Sprint went with regards to people, relationships, processes, and tools. It identifies and orders the major items that went well and potential improvements. It also creates a plan for implementing improvements to the way the Scrum Team does its work.

The Scrum Master encourages the Scrum Team to improve its development processes and practices to make it more effective and enjoyable for the next Sprint. The changes, however, must still occur within the Scrum process framework.

Does the Retrospective also review the product or work outcome?

No, the Product Increment is already reviewed by the Scrum Team and stakeholders in the Sprint Review. The Retrospective is about improving the way the product is built. It is not only the improvements of how Scrum is implemented by the team, but also the engineering practices like coding, testing, integration, deployment, etc.

During each Sprint Retrospective, the Scrum Team plans ways to increase product quality by adapting the Definition of Done as appropriate.

Outcome

By the end of the Sprint Retrospective, the Scrum Team should have identified improvements to implement in the next Sprint. Implementing these improvements is the adaptation to the inspection of the Scrum Team itself. Depending upon the effort involved, Scrum Team may resolve them immediately or add one or more to the Sprint Backlog of the next Sprint. Remember that it is optional to add them to Sprint Backlog.

Can a Scrum Team change fundamental of Scrum in the name of improvement?

The Scrum Team can make improvement plans on development processes and practices, but it cannot change the fundamentals of the Scrum framework itself.

-------------------Question- 42-------------------------

Which is true?

a) The Retrospective focuses on the Scrum Team's processes and people, and the Sprint Review focuses on the product.

b) The Retrospective focuses on the product, and the Sprint Review focuses on the Scrum Team's processes.

c) The Retrospective focuses on the Scrum Team's processes and people, and the Sprint Review focuses on velocity.

-------Answer-------

Correct answer is 'a.'

-------------------Question- 42-------------------------

Summary

- The Sprint Retrospective is a formal opportunity where the Scrum Team inspects itself and creates a plan for improvement for the next Sprint. At least one high priority improvement is identified to be added to the Sprint Backlog of the next Sprint.
- This is a three-hour time-boxed meeting for one-month Sprints. For shorter Sprints, the event is usually shorter.
- The Scrum Master encourages the Scrum Team to improve its development processes and practices.
- The Scrum Team plans ways to increase product quality by adapting the Definition of Done.
- The outcome is a list of most helpful improvements. The most impactful improvements are addressed as soon as possible. They may even be added to the Sprint Backlog for the next Sprint.

#

Chapter 5 - Controls for Scrum execution

--------------------------DE-TOUR--------------------------

Scrum does not explicitly define or list any controls. Yet, if you analyze Scrum, you can see that Scrum has many built-in risk controls, because Scrum is fundamentally a risk-reduction framework for addressing complex problems. This book features the prominent controls with the intention that Scrum users can appreciate the significance of such controls and try to fully leverage them in their Scrum implementations. The following controls minimize risk of producing waste and lack of progress:

1. Time-boxed Scrum events eliminate waste associated with traditional meetings.
2. Inspections increase the transparency of the Product Value and Work Progress.
3. The Scrum Master guards the Scrum Framework and champions the Scrum rules including transparency.
4. Standards like the Definition of Done set the expectations of the required quality levels and makes them transparent.

--------------------------DE-TOUR--------------------------

Time-boxed events

The Scrum events are significant controls in the hands of the Scrum Team that are used to maneuver the journey to the desired end.

- All events are time-boxed events, which means that every event has a maximum duration.
- The events create regularity and minimize the need for meetings not defined in Scrum.

The Scrum Master coaches the Scrum Team to correctly conduct and utilize the events. The Scrum Master need not be part of some of the events, but they need to ensure that these events happen.

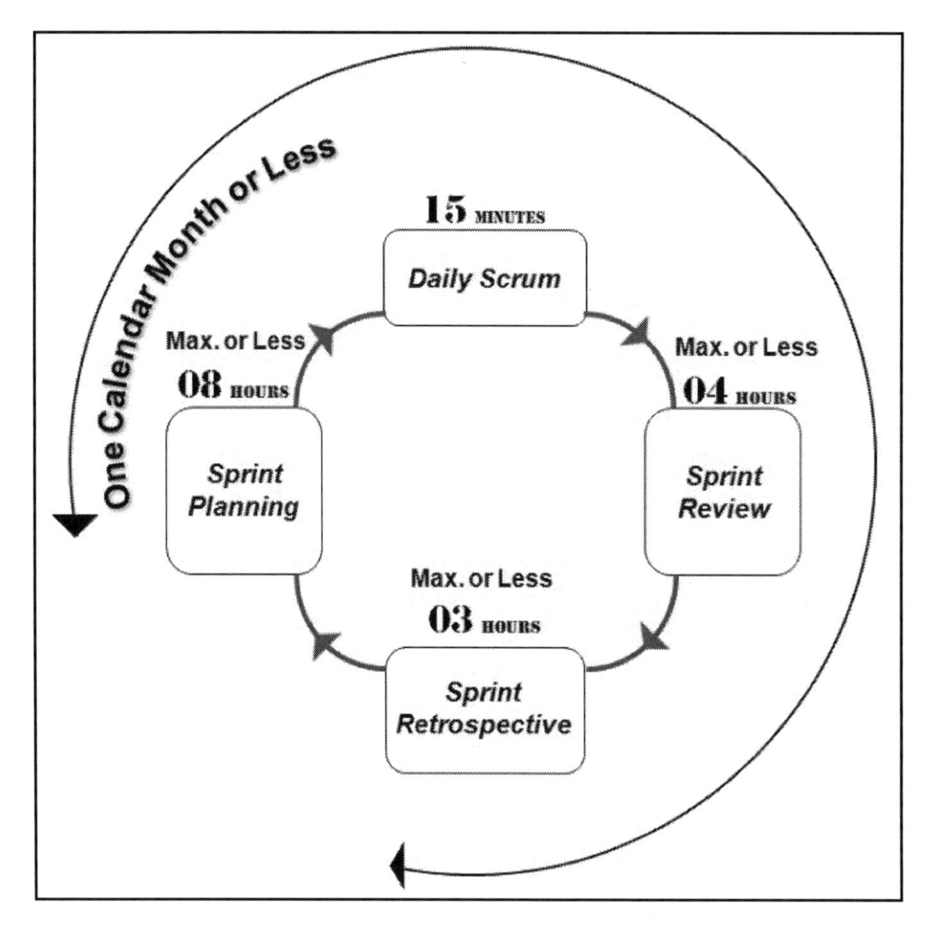

Fig. 17- A Sprint showing the time-boxed events

-------------------Question- 43-------------------------

A Scrum Team has only three Developers. The time box for the Daily Scrum is

a) 3 minutes

b) 15 minutes

c) Per team's decision

-------Answer-------

Daily Scrum is always time boxed to 15 minutes and scheduled that way. However, the team can close the Daily Scrum earlier on the days if they are done. Correct answer is 'b.'

-------------------Question- 43-------------------------

Once a Sprint begins, its duration is fixed and cannot be shortened or lengthened. The remaining events may end whenever the purpose of that event is achieved, ensuring an appropriate amount of time is spent without allowing waste in the process.

What happens if the Developers complete the work before the Sprint end?

If a team completes the work before the Sprint end, they may collaboratively decide to do additional work such as refining the Backlog, etc. and still complete the Sprint as planned. The Sprint duration is kept intact without contraction or expansion.

But for the other four events, Sprint Planning, Daily Scrum, Review, and Retrospective, if the team completes the intended agenda early, they can be closed early.

-------------------Question- 44-------------------------

Throughout the effort, who takes ownership of the Scrum events, sets-up the meeting for every event, and invites the required participants?

a) Product Owner

b) Scrum Master

c) Scrum Team

d) Developers

-------Answer-------

The Scrum Team is a self-managing team. They manage and organize how they perform their work and are collectively the owner of their work. The Scrum Team together comes up with the shared understanding of when to have these events. By bringing in this self-management and regularity, the team avoids the complexities of meeting arrangement and attendance associated with traditional meetings. The Scrum Master may facilitate this only during the early period but coach the Scrum Team to do it by themselves later. Correct answer is 'c.'

-------------------Question- 44-------------------------

Inspections

Every event is an opportunity for inspection. In addition, a team can optionally inspect more frequently. However, their inspection should not be so frequent that inspection gets in the way of the work.

Inspections are performed by those doing the work and those who have knowledge of the product. In addition to the Developers, the Product Owner and invited stakeholders inspect the product Increment during the Sprint Review.

If an inspector finds that any aspect of the work or product deviates from the acceptable limits and that the resulting product will be unacceptable, the work activity must be adjusted. The Developers do not wait for any formal event to make this adjustment; instead they make it as soon as possible to minimize further deviation.

-------------------Question- 45--------------------------

What should be the frequency of Inspection in Scrum? Select all that apply.

a) As planned in the Sprint Planning

b) As needed by the Product Owner

c) In every event within the Sprint

d) Frequently, as decided by the team, but not getting in the way of work

-------Answer-------

Every event is an opportunity for inspection. In addition, a team can optionally inspect more frequently, without having the inspections getting in the way of the work. Correct answers are 'c' and 'd.'

-------------------Question- 45--------------------------

Summary of Which Roles participate in Which events

Sprint Starts - Sprint Planning

Scrum Master	Product Owner	Developers	Technical/Domain Experts	Stakeholders (For e.g. sponsor)
Yes	Yes	Yes	Can be invited by Scrum Team	No

Ongoing Development Work

Scrum Master	Product Owner	Developers	Technical/Domain Experts	Stakeholders (For e.g. sponsor)
No (Yes, if they contribute to dev. work)	No (Yes, if they contribute to dev. work)	Yes	Can be involved** by Scrum Team	No

** In the middle of the Sprint, if the Developers need any external technical/domain help, they must raise this as an impediment, and they can reach out to technical/domain experts for help. However, later in the Sprint Retrospective, they need to inspect why they could not do it themselves without help. They need to identify the improvements to be made to make them truly cross functional and self-sufficient.

On a side note, if the Developers still cannot resolve an issue even with the external technical/domain help, they need to bring it up to the Scrum Master immediately and collaborate on the next course of action. If the Scrum Master cannot remove the impediment, then the Product Owner should be consulted.

Daily Scrum

Scrum Master	Product Owner	Developers	Technical/Domain Experts	Stakeholders (For e.g. sponsor)
Optional* (Yes, if they contribute to dev. work)	Optional (Yes, if they contribute to dev. work)	Yes	No	No

The Daily Scrum is an internal meeting for the Developers. If others are present, the Scrum Master ensures that they do not disrupt the meeting.

* The Scrum Master who also performs the Development Work must participate in the Daily Scrum. The Scrum Master who does not perform the Development Work can optionally participate to observe the correct implementation or facilitate discussions.

Ongoing Product Backlog Refinement

Scrum Master	Product Owner	Developers	Technical/Domain Experts	Stakeholders (For e.g. sponsor)
Optional	Yes	Yes	No	No

Sprint Review

Scrum Master	Product Owner	Developers	Technical/Domain Experts	Stakeholders (For e.g. sponsor)
Yes	Yes	Yes	Can be invited by Scrum Team	Can be invited by Product Owner

Sprint Retrospective – Sprint Ends

Only Scrum Team Participates				
Scrum Master	Product Owner	Developers	Technical/Domain Experts	Stakeholders (For e.g. sponsor)
Yes	Yes	Yes	No	No

-------------------Question- 46-------------------------

All the activities that happen within Scrum are called Scrum events.

a) True

b) False

-------Answer-------

There can be any number of activities within the Scrum Framework as chosen by the Scrum Team. It will include Development Work Practices, Backlog Refinement, etc. However, all the activities are NOT Scrum events. There are only five defined events in Scrum. Correct answer is 'b.'

-------------------Question- 46-------------------------

The Scrum Master's championship of Transparency

Scrum artifacts provide key information about the product and development work to both the Scrum Team and outsiders.

The artifacts are meant to maximize the transparency of the underlying information.

Scrum relies on transparency. Decisions to optimize value and control risk are made based on the perceived state of the artifacts. To the extent that transparency is complete, these decisions have a sound basis. To the extent that the artifacts are not transparent, these decisions can be flawed, value may diminish, and risk may increase.

The Scrum Master must work with the Product Owner, Developers, and other involved parties to understand if the artifacts are completely transparent. The Scrum Master's job is to work with the Scrum Team and the organization to increase the transparency of the artifacts. This work usually involves learning, convincing, and change. Transparency does not occur overnight but is a journey.

The Scrum Master and Artifacts

Note that the Scrum Master is not the owner of any artifact. However, the Scrum Master is the main driver in coaching the team to increase the transparency of these artifacts. The Scrum Master spends time in detecting incomplete transparency by inspecting the artifacts, sensing patterns, listening closely to what is being said, and detecting differences between expected and real results. There are practices for coping with incomplete transparency; the Scrum Master must help everyone apply the most appropriate practices in the absence of complete transparency.

-------------------Question- 47-------------------------

A good guideline to differentiate Acceptance Criteria from the Definition of Done is- the Definition of Done provides a checklist of quality measures to take the Increment to a potentially usable state, while the Acceptance Criteria focus on specifying the business features (functionality).

a) True

b) False

-------Answer-------

The Definition of Done is a commitment that defines the quality measures for a usable Increment (Fitness for use). Acceptance criteria are the specifications of the expected business features (Fitness for purpose). Correct answer is 'a.'

-------------------Question- 47-------------------------

The relationship between Roles and Artifacts and Events

The following table shows the relationship between the roles, events, artifacts, and commitments. It points out the creation and ownership of each Scrum artifact and the commitments.

Sprint Starts - Sprint Planning

Product Backlog	Sprint Backlog	Increment
Product Owner owns this. Product Backlog items that can be Done by the Scrum Team within one Sprint are deemed ready for selection in a Sprint Planning event	Developers own Sprint Backlog. In Sprint planning event, they create just enough Sprint Backlog to get work started	Developers create the Increment (after the Sprint Planning)
Commitment to Product Backlog: Product Goal	**Commitment to Sprint Backlog: Sprint Goal**	**Commitment to Increment: Definition of Done (DoD)**
The Product Goal is defined by the Product Owner and is in the Product Backlog. It describes a future state of the product which can serve as a target for the Scrum Team to plan against	Scrum Team crafts the Sprint Goal during the Sprint Planning event and then adds it to the Sprint Backlog	If an organization has existing DoD, Scrum Team takes it as minimum. Otherwise, Scrum Team creates it in first Sprint.

Daily Scrum

Product Backlog	Sprint Backlog	Increment
	Developers inspect progress toward the Sprint Goal and adapt the Sprint Backlog as necessary	

Ongoing Product Backlog Refinement

Product Backlog	Sprint Backlog	Increment
Product Owner breaks down (decomposes) and further defines Product Backlog items into smaller more precise items with Developers' help. Even outside this event, Product Owner may update Product Backlog anytime		

Sprint Review

Product Backlog	Sprint Backlog	Increment
Scrum Team and stakeholders review what (Increment) was accomplished in the Sprint and what has changed in their environment. The Product Backlog may also be adjusted to meet new opportunities		Scrum Team presents the Increment (Multiple Increments may be created within a Sprint). This Increment is additive to all prior Increments and is integrated so they work together
Commitment to Product Backlog: Product Goal	**Commitment to Sprint Backlog: Sprint Goal**	**Commitment to Increment: Definition of Done (DoD)**
Scrum Team presents the results of their work (Increment) to key stakeholders and progress toward the Product Goal is discussed		

Sprint Retrospective – Sprint Ends

Product Backlog	Sprint Backlog	Increment
	The most impactful improvements identified in the Retrospective may be added to the Sprint Backlog for the **next Sprint**	
Commitment to Product Backlog: Product Goal	**Commitment to Sprint Backlog: Sprint Goal**	**Commitment to Increment: Definition of Done (DoD)**
		The Scrum Team inspects how the last Sprint went with regards to individuals, interactions, processes, tools, and their DoD. DoD may be updated as needed

Summary

- Scrum has built-in controls to minimize the risk of producing waste and lack of progress.

- The first control is the Scrum events. They create regularity and minimize the need for meetings. All Scrum events are time-boxed events, such that every event has a maximum duration. The time boxing ensures that an appropriate amount of time is spent without allowing waste in the process.

- The second control is the Scrum Team Inspections. Other than the Sprint itself, which is a container for all other events, each event in Scrum is a formal opportunity to inspect and adapt. Inspections ensure that there is increased transparency of value and progress.

- The third control is the Scrum guardian, the Scrum Master. Since Scrum relies on Transparency, the Scrum Master must work with the Product Owner, Developers, and other involved parties to understand if the artifacts are completely transparent and must champion to increase the transparency of the artifacts by continuous learning, convincing, and bringing change.

- The fourth control is a commitment like the Definition of Done. Any product or system should have a Definition of Done that is standard for any work done on it. If the Definition of Done for an Increment is part of the conventions, standards or guidelines of the development organization, all Scrum Teams must follow it as a minimum. If there is no existing Definition of Done, the Scrum Team must define a Definition of Done appropriate for the product.

#

Chapter 6 - Closing Scrum

--------------------------DE-TOUR--------------------------

No information is found in The Scrum Guide about closing Scrum. This chapter addresses that for more clarity.

--------------------------DE-TOUR--------------------------

In Scrum, it is the Product Owner's call to decide the closure of the journey. Others can influence this decision. A Product Backlog lives as long as the associated product lives. A Scrum Team also lives as long as the associated Product Backlog lives.

Does the effort move from Development to Maintaining?

Along with features and functions, the Product Backlog can contain bug fixes too. There is no concept of 'moving the effort' from development to maintenance. The Scrum Team may be reduced in size, but there is no transition to a brand-new team.

When the Product is retired, the team may be released after the Sprint Review. In this case there is no need for a Sprint Retrospective as the team's work is over.

What happens if the work is performed by a service organization based on a contract?

In efforts that are performed by a Service Organization, the role of the Product Owner is performed by a person belonging to the Service Organization. This Product Owner and the Customer continuously collaborate and define the Product's needs in the Product Backlog. Any contractual needs like user documentation are also part of the Product Backlog. Rather than completing all the Product Backlog Items, achieving an appropriate value for the investment is the objective. Based on this, the Product Owner decides when to close the journey.

-------------------Question- 48--------------------------

After a Sprint Review, the Product Owner deems that the Product has come to the end of its life and the Product Backlog can be closed. The next immediate step is

a) To communicate the Scrum Team's availability to stakeholders.

b) To conduct a Retrospective.

c) To write transition documentation.

-------Answer-------

Usually the Retrospective is the last event of a Sprint. However, when the Product Owner decides that the development work is over, there is no need for a Retrospective. The transition documentation is defined as part of the Definition of Done if that transition documentation is a requirement for the Increment's release. A Done Increment would already have the transition document created. Correct answer is 'a.'

-------------------Question- 48--------------------------

Summary

This chapter is too short to review. Reread, if necessary.

#

Quiz 3

1. In a Retrospective, a Scrum Team decides to revise the Sprint length. The new Sprint length needs to be agreed upon by the Product Owner.

a) True

b) False

-------Answer-------

The Product Owner needs to ensure that the Sprint length is short enough to limit business risks and also short enough so the team can synchronize the development work with other business events. So, it requires the approval of the Product Owner. Please note that the finalized Sprint length cannot be longer than 1 calendar month. Correct answer is 'a.'

2. In the middle of the Sprint, the Developers find that some of the Product Backlog Items forecast for this Sprint cannot be finished because they need significant additional effort. However, the Developers can still meet the Sprint Goal with rest of the items. The next thing to do is

a) Consult with the Product Owner. If they agree to cancel the current Sprint, plan a new Sprint with new estimates.

b) Do not cancel the Sprint. Extend the Sprint duration as required for the additional effort.

c) Collaborate with the Product Owner and negotiate the removal of the Product Backlog Items that cannot be finished. Add new items prioritized by the Product Owner up to team's capacity. Complete the Sprint.

-------Answer-------

As a first step, the team needs to solve this on its own. If they cannot, they should capture this as an impediment and try to work with the Scrum Master. After that, if the impediment is not solved, they need to involve the Product Owner. Correct answer is 'c.'

3. The Scrum Team gathers for the Sprint Planning meeting. The Product Owner has some stories, but the team finds that the stories do not provide enough information to make a forecast. The next immediate step is

a) The Scrum Master cancels the Sprint.

b) The Developers proceed with starting with whatever is known.

c) The Developers make it transparent that they cannot make a forecast with insufficient information and negotiates with the Product Owner on refining the stories to a ready state.

d) The Scrum Team discusses the root cause in the Retrospective.

-------Answer-------

The Product Owner needs to help clarify the selected Product Backlog Items. The Scrum Master can also coach the Product Owner on how to accomplish this. One example is by having regular "backlog refinement sessions." Answer 'd' is also correct, but the question asks about the "next immediate step." Correct answer is 'c.'

4. Conducting the Daily Scrum at the same time and same place every day makes it easier for the Product Owner and the Scrum Master to participate.

a) True

b) False

-------Answer-------

It is preferable to conduct the Daily Scrum at the same time and same place so it can reduce the complexity of meeting overhead. Unless the Product Owner and Scrum Master are also performing Developer activities part time, they are not required for the Daily Scrum. Correct answer is 'b.'

5. A discussion of what to do next is an additional event in the Sprint Review.

a) False

b) True, and the scope of the next Sprint is also finalized here.

c) True, and it may capture probable Backlog Items for the next Sprint, but the scope of the next Sprint is deferred until Sprint Planning.

-------Answer-------

Every Scrum event is an opportunity for inspection and adaptation. In the Sprint Review, inspecting the product Increment provides insights and clarity. This newly found knowledge is used to 'adapt the next steps,' i.e., find out what to do next. Correct answer is 'c.'

6. Given a complex product and its relevance to multiple departments, a Scrum Team expects that they need to invite many stakeholders for the Sprint Review. It estimates that the review will take more than 4 hours. The Scrum Team can increase the Sprint Review duration.

a) True

b) False

-------Answer-------

Every Scrum event is time-boxed to reduce risk and eliminate waste. If there are valid reasons that require more time, it might be due to reasons such as added complexity. A team needs to address these root causes, and the time box should be followed to contain risks. Correct answer is 'b.'

More examples of where the time box is likely to be broken but should be safeguarded are

- A Daily Scrum cannot exceed 15 minutes even if the team size is increased.
- A Sprint cannot go beyond one month even if there are items almost done and may need just few more days to be completely Done.

7. A Scrum Team maintains a Sprint burn-down to track estimated remaining work. In the middle of the Sprint, the burn down graph shows an upward spike. This indicates

a) A planned work is removed.

b) The Product Owner added a new item to the Sprint.

c) The Developers added new work.

-------Answer-------

A spike indicates added work. The Product Owner cannot add new work without the Developers' consent. Correct answer is 'c.'

8. The value attached to the Product Backlog Item is guaranteed to be realized.

a) True

b) False

-------Answer-------

The value is an estimate based on assumptions. It has to be validated by releasing the item. Scrum facilitates the early validation by making it available as a valuable and useful Increment to the Product Owner who can then choose to release it to Production. Correct answer is ‘b.’

9. Who finalizes the number of Product Backlog Items that can be selected for the Sprint Backlog?

a) The Product Owner since they maximize the value of work of the Developers.

b) The Scrum Master since they coach the team on Scrum.

c) The Developers since they are the owners of the work.

d) The Scrum Team together negotiates and reaches an agreement. They may use the team velocity as a standard to calculate how much work they can take.

-------Answer-------

The Product Owner optimizes the team’s work by keeping the Product Backlog ordered, and hence deciding what they work on next. Only the Developers can finalize how many Product Backlog Items it can complete in the Sprint. Correct answer is ‘c.’

10. After Sprint Planning, the Product Backlog Items selected into the Sprint Backlog are frozen and cannot be modified. The only way to modify the Sprint Backlog is to have the Product Owner cancel the Sprint.

a) True

b) False

-------Answer-------

The Sprint Goal gives the Developers some flexibility regarding the functionality implemented within the Sprint. As the Developers work, they keep the Sprint Goal in mind. If the work turns out to be different than what the Developers expected, they collaborate with the Product Owner to negotiate the scope of the Sprint Backlog within the Sprint. Correct answer is ‘b.’

#

Part 3 – More PSM Exam specific material

"The greater danger for most of us lies not in setting our aim too high and falling short; but in setting our aim too low, and achieving our mark." – Michelangelo [Italian sculptor, painter, architect, poet, and engineer of the High Renaissance...]

#

Chapter 1 - Detailed view of the Assessment process

#

Chapter 1.1 - Overview of all Scrum.org Assessments

The following are the different assessments to get certifications. Once acquired, the certificates need NOT be renewed again. Please verify any information related to the assessments by checking www.scrum.org. You can also send your questions to the friendly and super responsive support group at support@scrum.org.

Scrum Master and Anyone willing to learn Scrum

For assessing Scrum or Scrum Master knowledge, there are four levels of assessments.

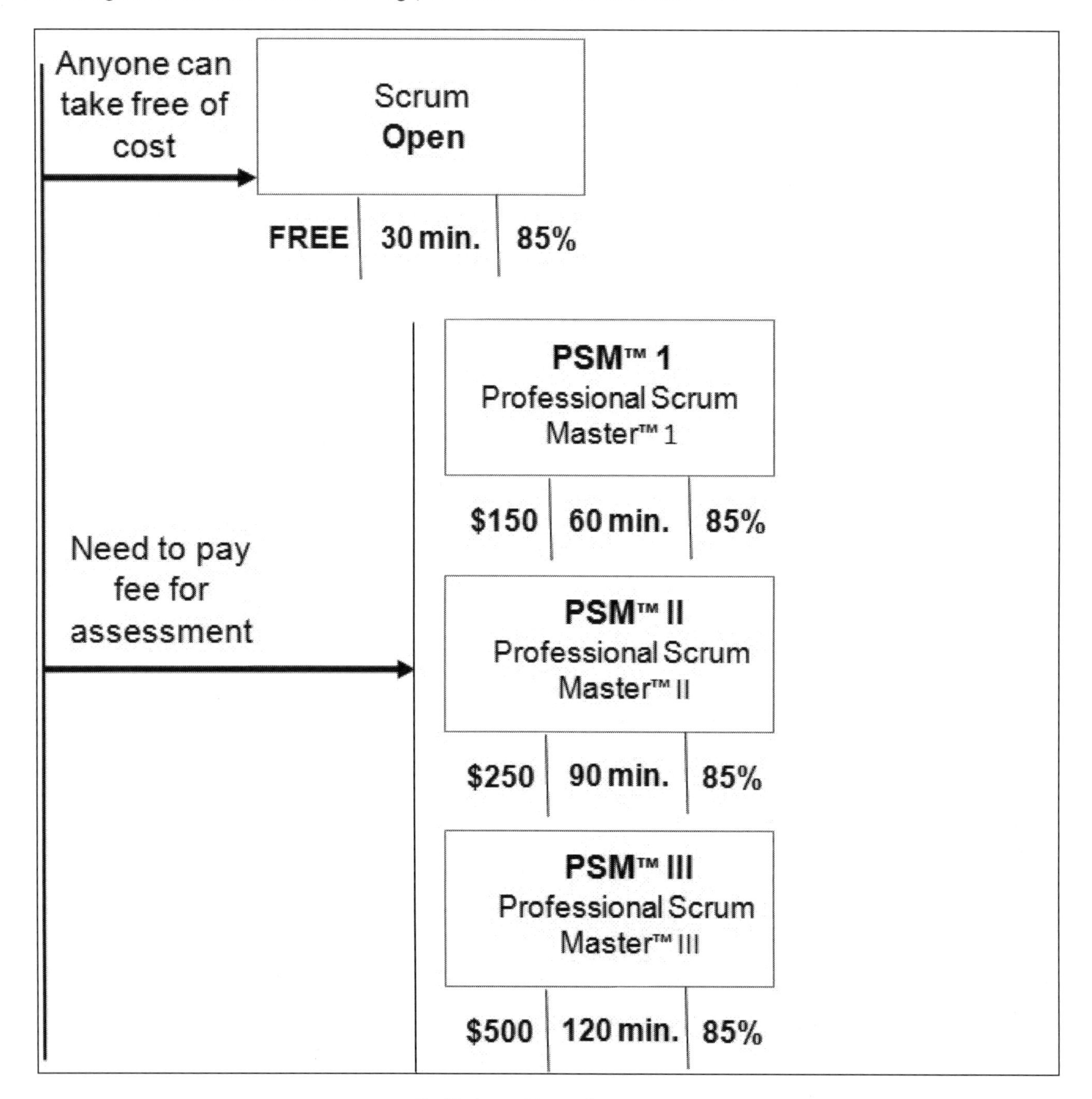

Fig. 18- Scrum Master Assessments

Scrum Open

- Tests the basic knowledge of Scrum.
- Offered free of cost and can be taken by anyone by accessing the Open-Assessments at www.scrum.org/.
- Consists of 30 questions with a 30 minute time limit.
- There are no limits on the number of attempts.
- This free assessment also provides an idea of the type and structure of the questions that are asked in the paid Professional Scrum Master 1 (PSM 1) assessment. Before taking the PSM 1, it is highly useful to practice using the free Scrum Open.

Professional Scrum Master™1 (PSM 1)

- Tests the intermediate knowledge of Scrum.
- Requires a payment of $150 and can be taken from anywhere by buying the assessment at www.scrum.org/Assessments/Professional-Scrum-Master-Assessments/PSM-I-Assessment. Click on the "BUY PSM 1 ASSESSMENT" button.
- Consists of 80 questions with a 60 minute time limit.
- Only one attempt is allowed.
- The subject of this book is the preparation for this assessment.

Professional Scrum Master™ II (PSM II)

- Tests the advanced knowledge of Scrum.
- Requires a fee payment of $250 and can be taken from anywhere.
- Consists of 30 questions with a 90 minute time limit.
- Only one attempt is allowed.
- Passing the PSM 1 is mandatory before attempting the PSM II.

Professional Scrum Master™ III (PSM III)

- Tests the in-depth knowledge of Scrum.
- Requires a fee payment of $500 and can be taken from anywhere.
- Consists of multiple choice and essay-type questions with a 120 minute time limit.
- Only one attempt is allowed.
- Passing the PSM II is mandatory before attempting the PSM III.

Scrum Product Owner

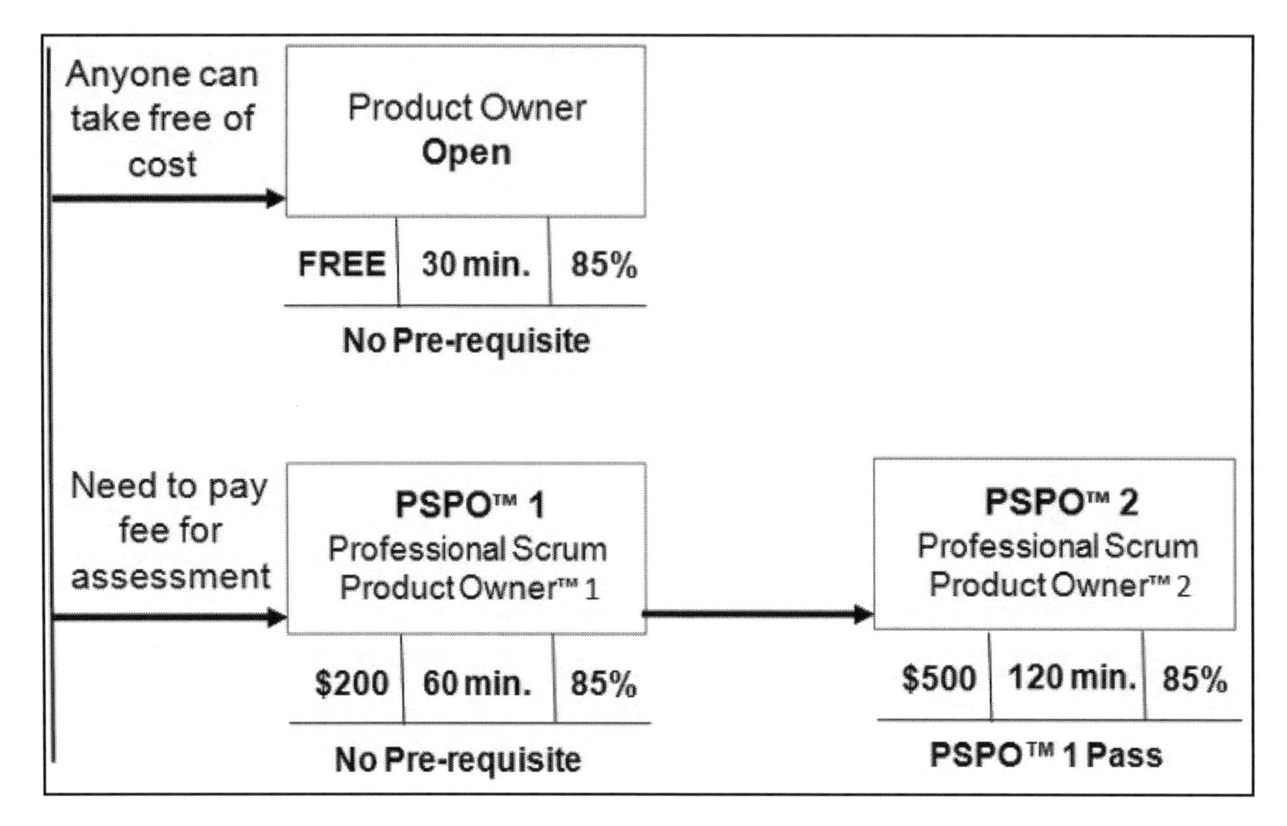

Fig. 19- Product Owner Assessments

Product Owner Open

- Tests the basic knowledge of Scrum needed for the Product Owner role.
- Offered free of cost and can be taken by anyone.
- Consists of 15 questions with a 30 minute time limit.
- There are no limits on the number of attempts.
- This free assessment also provides an idea of the type and structure of the questions that are asked in the paid Professional Scrum Product Owner 1 (PSPO 1) assessment. Before taking the PSPO 1, it is highly useful to practice using the free Product Owner Open.

Professional Scrum Product Owner™ I (PSPO I):

- Tests the intermediate knowledge of the Product Ownership aspect in Scrum.
- Requires a fee payment of $200 and can be taken from anywhere.
- Consists of 80 questions with a 60 minute time limit.
- Only one attempt is allowed.

Professional Scrum Product Owner™ II (PSPO II):

- Tests the advanced knowledge of the Product Ownership aspect in Scrum.

- Requires a fee payment of $500 and can be taken from anywhere.
- Consists of multiple choice and essay-type questions with a 120 minute time limit.
- Only one attempt is allowed.
- Passing the PSPO I is mandatory before attempting the PSPO II.

Scrum Developer

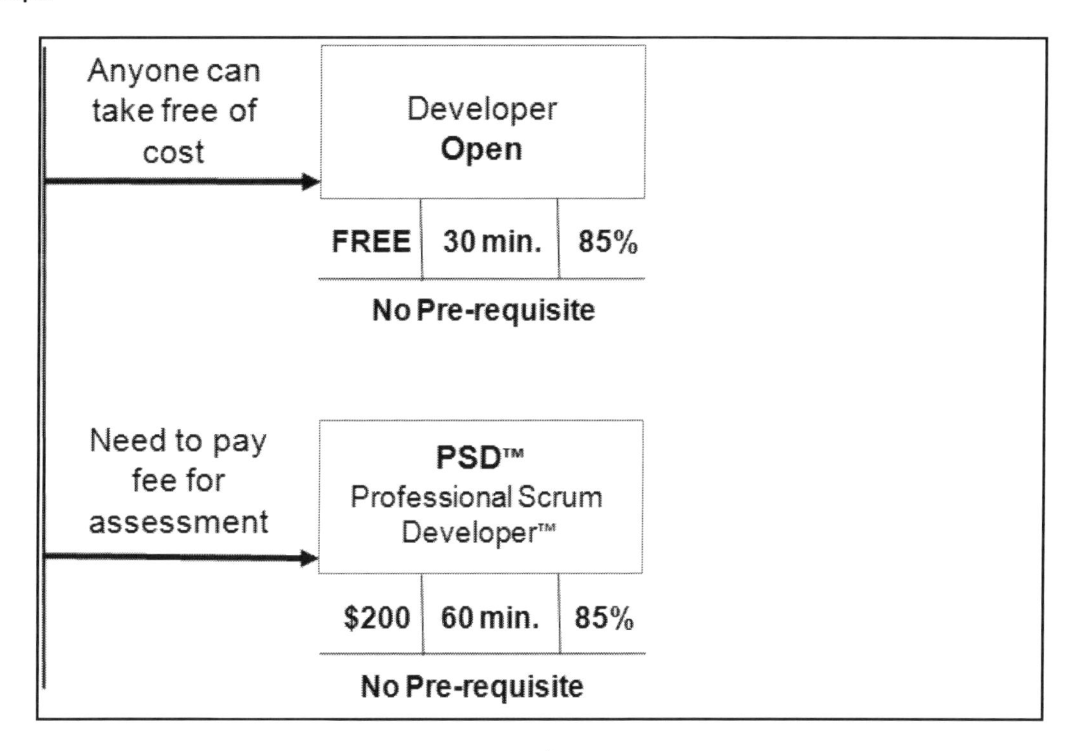

Fig. 20- Scrum Developer Assessments

Developer Open

- Tests the basic knowledge of Scrum needed for the Developer role.
- Offered free of cost and can be taken by anyone.
- Consists of 30 questions with a 30 minute time limit.
- There are no limits on the number of attempts.
- This free assessment also provides an idea of the type and structure of the questions that are asked in the paid Professional Scrum Developer (PSD) assessment. Before taking the PSD, it is highly useful to practice using the free Developer Open.

Professional Scrum Developer™ (PSD):

- Tests the software engineering principles like Test First Development, Continuous Integration, etc. within the Scrum framework.
- Requires a fee payment of $200 and can be taken from anywhere.
- Consists of 80 questions with a 60 minute time limit.
- Only one attempt is allowed.

Other Assessments for Experts – Scaled Scrum

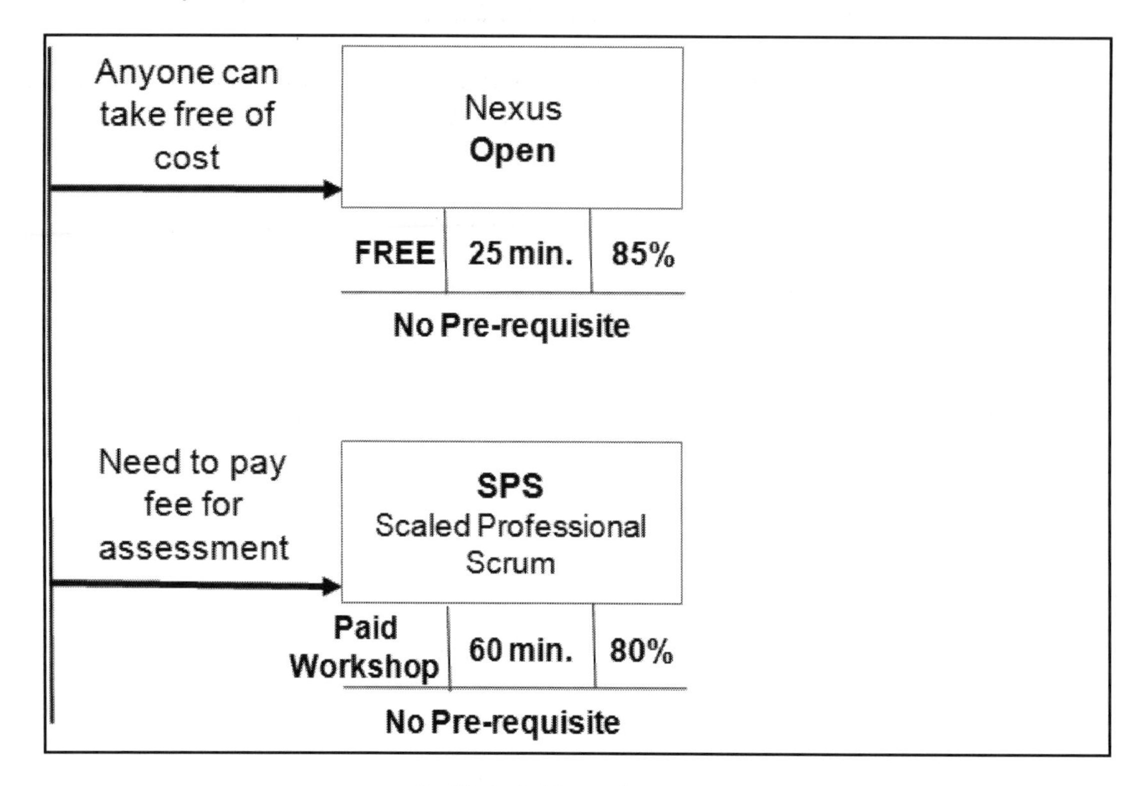

Fig. 21- Scaled Scrum Assessments

Nexus Open

- Offered free of cost and can be taken by anyone.
- Consists of 15 questions with 25 minute time limit.
- There are no limits on the number of attempts.

Scaled Professional Scrum

- Tests the knowledge of scaling Scrum for large software development projects.
- This is a professional assessment that requires taking a paid training course by Scrum.org and attempting the assessment after the training course.

------------------Question (Not a potential question in PSM assessment)----------------

It is mandatory to take a training course and pass the Scrum Open Assessment before taking the PSM 1 assessment.

a) True

b) False

-------Answer-------

The Scrum Open is a free online assessment for anyone to test their fundamental knowledge of Scrum. Though it is highly useful as a practice assessment, it is not mandatory before any assessment. Correct answer is 'b.'

------------------Question (Not a potential question in PSM assessment) -----------------

#

Chapter 1.2 - Professional Scrum Master™ 1

The PSM 1 tests the demonstration of intermediate knowledge of Scrum. It includes Theory of Empiricism, Lean Thinking, Team Self-Management, Servant Leadership and Coaching, and Scrum Teams and their associated roles, events, artifacts, and rules. Some knowledge of optional techniques of Scrum also may be needed.

It is noteworthy to mention that the knowledge verified through PSM 1 will help in other ways too:

- Passing the PSM 1 is mandatory before attempting PSM II: Even for other assessments such as PSD and PSPO, one needs to have foundational knowledge of Scrum that is tested in the PSM 1.

- Also, the PSM 1 is the first step towards acquiring the Professional Scrum Trainer (PST): To apply for PST one needs to get 95% or higher in the PSM 1 assessment. If one has passed the assessment but failed to get 95%, they can retake the assessment until they get the required percentage.

-------------------Question (Not a potential question in PSM assessment)-----------------

The PSM 1 assessment tests

a) The job skills of a Scrum Master.

b) The understanding of engineering practices of Scrum.

c) The intermediate knowledge of Scrum.

d) The Scrum knowledge through Scrum Open.

-------Answer-------

Correct answer is 'c.'

-------------------Question (Not a potential question in PSM assessment)-----------------

Certification Qualification and Format

- There are no prerequisite qualifications for this assessment.

- It consists of 80 multiple choice questions that include

a) Questions with one correct answer.

b) Questions with multiple correct answers. To get full points for these questions, all correct answers must be selected. There is no point awarded for partially correct answers.

c) Questions answered with Yes or No and Questions answered with True or False.

- This assessment is offered only in English. Non-native speakers without English as a professional language may have minor issues in understanding some of the questions.

- The time limit is 60 minutes.

- To pass one needs to score 85% or higher. It translates to 68 questions answered correctly, but sometimes a question can be awarded a weight other than 1 point. How much weight each question carries is not made explicit.

- You can bookmark a question without answering it and move on to the next question. You can go back to bookmarked questions during the 60 minutes.

Certification Fee, Payment Mode, and Registration

- Please check Scrum.org for the latest fee. As of March 2021, it is $150.

- If you took a paid training class from Scrum.org like Professional Scrum Foundations™ or Professional Scrum Master™, you would get two free attempts at the PSM 1. The password for the assessment will be emailed to within 3-5 business days after successfully completing the class. However, taking a course is not required to take the assessment.

- To register for the assessment, go to https://www.scrum.org/professional-scrum-master-i-certification. Click on the "BUY PSM 1 ASSESSMENT" button.

- You can use credit/debit card to pay.

- If the card is non-USA, the equivalent amount for 150 US dollars will be deducted by your card issuer subject to their currency conversion rules. Sometimes you have to call your card issuer (bank) to remove any limits/security settings on your card before paying. You can also clarify with them about the local currency conversion rate that will be applied.

- After paying the fee, the password for the assessment is sent to the registered email usually within a business day.

- The password does not have an expiration date.

- Only one attempt is allowed.

Certification Medium – No Test Center

- Scrum.org assessment's including the PSM 1 can be taken from your place of choice. You do not need to go to a Test Center.

- In other words, it is an open book test. However, you cannot depend on just looking online or in The Scrum Guide to find the answers. The questions may not be so straightforward.

- You will need a computer with a reasonably fast internet connection.

- The online assessment works on any web browser: Internet Explorer, Chrome, and Safari.

Taking the Assessment

- You can access Scrum.org and use the password to take the assessment.

- In case the connection is lost, or the online assessment terminates abruptly due to technical reasons, you can contact support@Scrum.org with evidence. They may offer a discount for the next attempt, but the author recommends you clarify with them if you want to confirm or clarify any questions.

- With sufficient preparation using this book, you can complete most of the questions within 45 minutes. This will provide 15 minutes more to go back to bookmarked questions. Using this book, you can further increase your speed by training yourself to answer each question within 30 seconds.

After the Assessment – Passing and Certificate

- At the earliest occurrence of submitting the quiz or the 60 minutes ending, the results will be shown on the screen. If you scored 85% or higher, you would see a congratulatory message on the screen. The breakdown of the results will look something like this

Scrum Framework - XX%

Theory & Principles - XX %

Cross-Functional, Self-Management - XX%

Coaching & Facilitation - XX%

- You will get the PSM 1 certificate in an email in 5 - 7 business days.
- After passing the assessment, your name will be posted on Scrum.org under the PSM 1 Certificate Holders list.
- **Once acquired the certificate need not be renewed.**

After the Assessment – Not Qualifying

If you scored less than 85%, you can write to Scrum.org to get feedback on specific areas of improvement.

Scrum.org will not provide the details of exact questions that went wrong. However, you will know the concepts around which you had difficulty.

An example of feedback for someone who had difficulty in understanding the concept of self-management is provided below. This is only indicative of the kind of feedback and is not an authorized version from Scrum.org.

'You got a number of questions incorrect around Scrum roles and how they support Scrum principles. Focus on understanding the roles within Scrum. If you face a question that asks what a Scrum Master would do in a particular situation, think about their role. The Scrum Master teaches the team to solve the problems themselves using Scrum values. Your previous experience as a leader may push you to own the problem of others and solve it. However, it is against building up a self-managing team. A Scrum Master does not serve the team if they tend to own something and do it when the team can own and do it themselves.'

Chapter 2 - How to Prepare for the PSM 1 Certification

Before looking at the approach to prepare for the assessment be assured that one can clear the assessment without practical job knowledge of Scrum.

This book helps you to get the required information and practice as well.

Use this book to Learn, Prepare, and Practice

You can follow the approach recommended in Figure 22.

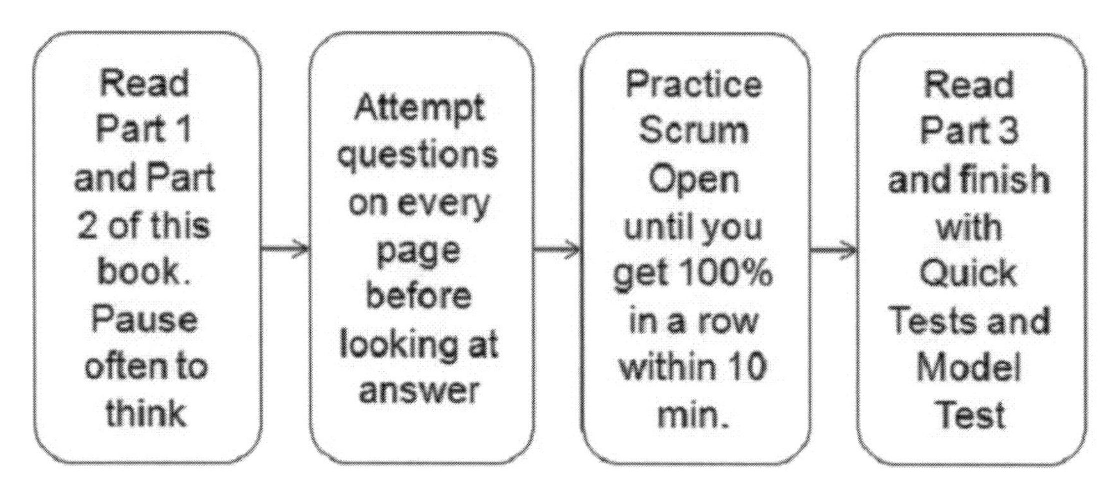

Fig. 22- PSM Preparation Approach

Understand Scrum and Practice for the PSM 1 at the same time

In general, it is important to understand absolutely everything in the short "The Scrum Guide." It is available free of cost at http://www.scrumguides.org/. This book explains The Scrum Guide.

While studying Scrum try to answer the questions that are on almost every page of this book. The questions are provided immediately after introducing any concept. These questions make you think about the granular interpretations behind every statement of The Scrum Guide. These granular interpretations are usually not evident in normal reading, but pop out when backed by questions. Such learning helps in a firm anchoring of what you just read.

The questions in this book represent questions that may be asked on the PSM 1 test. So, it allows you to practice for the assessment too.

Fully Prepare for the PSM 1 from all angles:

There is lot of additional contextual information that not only makes you fully prepared for the assessment, but also aids your job skills. For example, this book helps you develop the understanding how Scrum works in larger organizational ecosystems by pointing out peripheral interfaces of Scrum like management, stakeholders, etc. At specific places this book highlights some techniques like burn-down that are usually employed by practitioners in an actual situation.

For the experienced Scrum practitioners, there is some amount of unlearning and clearing of misconceptions required, so they can correctly understand the PSM 1 questions and appropriate answers from an authentic Scrum perspective. This book has a separate chapter on that.

Finally, use this book to know useful information and tips such as word play in Scrum, Agile versus Scrum, etc. so that you are covered from all angles.

Benchmark with Scrum Open Assessment

Take the Scrum Open assessment which is available freely to anyone online at www.Scrum.org/Assessments/Scrum-Open-Assessment. For each question, do not just stop with what was the right or best choice. Look at all the choices and clearly understand why they are not the best choices.

If you score more than 85% when taking the Scrum Open assessment, then you have good understanding of the assessment.

When you repeat the Scrum Open assessment, many of the same questions repeat. So, after taking the assessment the first time, set a time limit of less than 10 minutes to complete the Scrum Open assessment. DO NOT EVER RELAX THIS TIME. If you really understand Scrum, you will think faster. Verify that you are able to score more than 90% within 10 minutes.

Make multiple attempts at the Scrum Open assessment until you get 100% in less than 10 minutes several times in a row.

More Practice

This book provides more than 250 PSM 1 assessment-related questions. In Part 3, there are five quick tests with ten questions each. If you score high in the first test, i.e., a minimum of 8, then you can go to the next quick test. Finally, you can take one complete test which simulates an actual PSM 1 assessment. If you can get through that with a score of 85% or higher, you are well prepared.

Last day before the Assessment

The summary behind each chapter is a nugget of Scrum. Revisit those. Also, many chapters are designed in question and answer format. You can have someone ask these questions, and you can ensure that you know the answers.

Question Types

The PSM 1 assessment examines the Scrum knowledge in the following subject areas:

- Scrum Theory and Principles
- The Scrum Framework
- Coaching and Facilitation
- Cross-functional and Self-Managing teams

The following section provides a detailed idea about the type of questions that may be asked in the assessment along with examples and how this book helps with each type.

Question directly from the Open Assessment

From the feedback of multiple PSM 1 holders, 5 – 10% of the questions in the actual assessment are directly from the Scrum Open assessment. That's also another reason to attempt the PSM Open assessment until you get 100%.

How much does this book cover these types of questions? Getting 100% in a PSM 1 Open assessment is enough. Knowledge of Part 1 and Part 2 will prepare for the Open Assessment.

Question directly from The Scrum Guide

These are questions that are straightforward but require you to remember every word of Scrum. Active learning helps to remember the core content. First, study this book. Then, read The Scrum Guide word by word.

How much does this book cover these types of questions? Actively learning Scrum using this book will cover all the questions directly from The Scrum Guide.

-------------------Question- 1-------------------------

Having more than ten members on a Scrum Team

a) Is good because a larger team increases productivity.

b) Is good because there is more opportunity for cross training and backups.

c) Is not recommended because smaller teams communicate better and are more productive.

d) Is not recommended because the Scrum roles cannot be provided to everyone on the team.

-------Answer-------

The source of the answer is a direct statement from The Scrum Guide: "*The Scrum Team is small enough to remain nimble and large enough to complete significant work within a Sprint, typically 10 or fewer people. In general, we have found that smaller teams communicate better and are more productive.*" Correct answer is 'c.'

-------------------Question- 1-------------------------

Questions outside The Scrum Guide

The following is a sample of the question that you cannot answer directly from The Scrum Guide. To answer this specific question, you need to know a technique that is not explained in The Scrum Guide. It is expected that you as a Scrum Master with some practical experience know certain "real-time techniques." This book covers enough of these techniques where the questions can be anticipated.

<u>How much does this book cover these types of questions?</u> Thorough understanding of this book is essential. Attempting the questions on your own without jumping into the answers will provide an opportunity to think through and develop such understanding.

-------------------Question- 2-------------------------

The Scrum Team is in the middle of a Sprint. The burn-down indicates that there is a big divergence between planned burn-down and actual burn-down. The inference is

a) The Scrum Master did not plan the Sprint properly.

b) There is more remaining work to do than originally anticipated.

c) There is less remaining work to do than originally anticipated.

d) The Developers need to re-plan as soon as possible.

-------Answer-------

The actual progress is different from what was forecast by the team. So, the team has to re-plan to meet the Sprint Goal. Other answers are incorrect because though there is a divergence, there is no indication if the team is ahead or behind. Also, the Scrum Master is not the owner of the planning.

The Scrum Guide touches upon burn-down on a fleeting note only with no description. Answering this question requires additional knowledge beyond that description. Correct answer is 'd.'

-------------------Question- 2-------------------------

Questions that check the Deeper Meaning and Experience

This is the difficult type of question. Such questions require a good understanding of the Scrum concepts around Self-management, Servant Leadership, Product Ownership, handling issues at the periphery of the Scrum Team's boundary, etc. and some practical experience of applying Scrum. This book addresses these needs.

-------------------Question- 3-------------------------

For a Scrum Team, the Sprint Planning meetings are always going beyond the time-boxing. What could be the likely causes? Select all that apply.

a) The Scrum Master does not moderate and control the participants.

b) The Team didn't invest enough into Backlog Refinement.

c) The Product Backlog size is huge.

d) The Developers are trying to get a perfect and detailed Sprint plan.

-------Answer-------

The Scrum Master's role is not to control people or discussions but let the Team self-manage. They only coach and educate the Team to become self-managing. The Product Backlog size does not impact the time because the team does not need to discuss all items in the Product Backlog only those that are ordered on the top and are sufficiently deemed "ready" to be pulled into the Sprint.

Most teams are usually stuck with Product Backlog Items that are not decomposed and refined to a level that have sufficient clarity and transparency so they can be done within a Sprint. If the Team has not continuously engaged in Backlog Refinement sessions, they will end up doing "Just in Time refinement" during Sprint Planning.

The chosen Product Backlog Items and the details of work planned for first few days of the Sprint are enough to close the Sprint Planning and start the work. The Developers do not need to create a detailed work plan for a complete Sprint in the Sprint Planning. They can update the work plan as more details emerge during the Sprint.

Correct answers are 'b' and 'd.'

-------------------Question- 3-------------------------

#

Chapter 3 - Experienced Practitioners – Is it original Scrum?

Many Scrum practitioners acquire their knowledge of Scrum from multiple sources - some on the job, some from the open literature, some from various training sessions, and some with a combination of a few or all of these.

By virtue of their experience from multiple sources, they develop their "own understanding and interpretation" of what Scrum is including the roles, events, artifacts, and rules.

Important:

If you acquired your Scrum knowledge from random sources (writings / talks / trainings that are not authentic), please read the following statement very carefully.
For PSM testing purposes, the approved body of knowledge of Scrum is 'The Scrum Guide' authored by Jeff Sutherland and Ken Schwaber. Scrum is 'The Scrum Guide,' available at https://scrumguides.org/

Scrum - Intentional Flexibility and Binding the Optional Techniques

Scrum clearly defines each event, role, artifact, rule, and those that constitute the Scrum framework. However, there are many sub-techniques of product development that are left undefined intentionally.

While it is optional to apply the technique that works for the team, Scrum does not bind the Scrum users to specific sub-techniques. The Scrum Team is expected to figure out the best methods, techniques, and practices themselves. Scrum is lightweight and can be implemented without the need of sub-techniques. Some of the experienced Scrum practitioners working in an environment where teams "filled" these intentional gaps with some techniques tend to associate these techniques with Scrum. Here is an indicative list of those that are not prescribed by Scrum but are often mistaken as a definitive part of Scrum.

Writing Product Backlog Items as User Stories:

Scrum does not prescribe a specific format for how the Product Backlog Items (requirements) are defined. Many Scrum Teams use the format of **User Stories** to define the Product Backlog Items, and find them useful. A User Story format looks something like the following

As a <User Role>, I would like <this feature / action>, so that I can <specific benefits>.

------------------Question- 4-------------------------

A Scrum Team is in the process of defining the Product Backlog Items. The Scrum Master notices that the team is not using the User Story format to capture the Product Backlog Items. The Scrum Master should

a) Correct the team's behavior by coaching them about User Stories.

b) Add a business analyst with the knowledge of writing User Stories to the team with the specific responsibility of documenting the Product Backlog with User Stories.

c) Let the team decide the format of the Product Backlog Items.

-------Answer-------

Scrum does not define a specific technique for documenting the Product Backlog Items. Correct answer is 'c.'

-------------------Question- 4-------------------------

Applying Specific Estimation Techniques

Scrum does not prescribe a specific estimation technique. Often the practice of estimation by the Planning Poker card game is associated with Scrum. Planning Poker is a practice where the Developers along with the Product Owner estimate the Product Backlog Items as described below.

1. Each Developer holds a set of plain index cards.

2. The Product Owner introduces a Product Backlog Item and clarifies any questions.

3. Now each team member writes the estimate for this Item on one of the index cards without revealing the estimate to anyone else.

4. Then all the team members reveal their card with their estimate at the same time. If there are only marginal differences between their estimates, they arrive at a consensus by discussion.

5. If there are outliers by considerable margin, the owners of those outliers explain how and why they came up with those estimates. This may lead to more discussion and questions for the Product Owner increasing the overall clarity.

6. Then the team goes for the next round of estimation. They repeat this process until the team reaches complete consensus about all the estimates.

-------------------Question- 5-------------------------

The estimation method recommended by Scrum is

a) Poker Game.

b) T-Shirt Sizing.

c) Expert Judgement.

d) None of the above.

-------Answer-------

Any technique that is useful can be chosen by the Developers. Correct answer is 'd.'

-------------------Question- 5-------------------------

Using Task boards / Kanban boards / Scrum boards to visually communicate the artifacts:

These are the visual representations of the Sprint Backlog and the current work status. These boards are useful in making the information publicly available on the product development floor. Also, they can be used to increase the efficiency of managing the work items through the **lean** principles like 'Flow', 'Limit the Work in Progress', etc. Lean is another powerful philosophy of product development like Agile.

Though many Scrum Teams may use this, the task board or Kanban board is an optional implementation in Scrum.

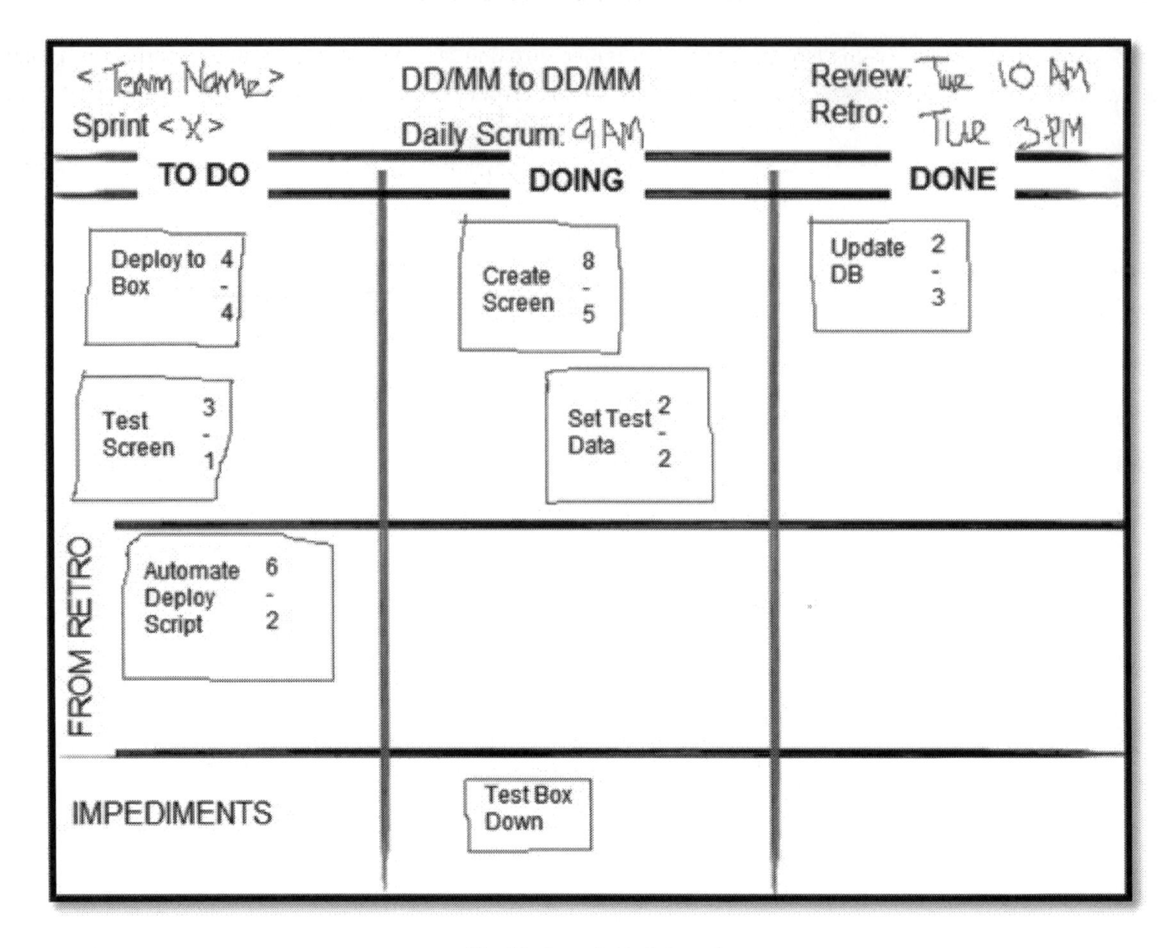

Fig. 23- Sample Task Board

-------------------Question- 6-------------------------

A Scrum Team has created the Sprint Backlog in the form of a task board. What is your inference?

a) The team can choose to represent the Sprint Backlog in any form that makes sense for them.

b) It is okay to have it in task board format, but it must be ensured that it follows all the Kanban principles.

c) The Scrum Master must advise the team to create a proper Sprint Backlog in the form of a matrix of the selected Product Backlog Items, related tasks, estimations, owners, and expected completion dates.

-------Answer-------

It is up to team to choose the format of the Sprint Backlog. Correct answer is 'a.'

-------------------Question- 6-------------------------

Tracking technical debt

In software development, technical debt refers to the sub-standard technical work that leaves gaps in the technical quality of the Product Increment. A Product Increment that is already tested to be functionally fit and useable in the market can still have bad technical quality and hence the technical debt. Technical debt accrues due to bad technical choices. Technical debt will need additional work to improve the quality of the Product Increment and may increase the cost of the maintenance of the product.

An example of technical debt is – writing lengthy code, all dumped into one file, leading to several thousand lines of code in that single file. Such code will look unreadable, be prone to injecting errors while introducing the code changes later, increase the dependencies during code deployments, and so on. In this example, an acceptable standard would be to write simple and short code in modules. To fix this technical debt (bad technical choice of lengthy code), code refactoring (modifying to short code modules) is needed later.

There are tools that scan the software code and come up with a numerical index that reflects the technical debt of that code. Many Scrum Teams measure technical debt, but there is no specification of technical debt within Scrum. However, it is useful to know what the technical debt is.

-------------------Question- 7-------------------------

Effort required to fix or refactor a product after it has been built is known as

a) Maintenance.

b) Technical Debt.

c) Plumbing code.

-------Answer-------

Technical debt is accrued as a result of making poor technical choices. Technical debt is not part of Scrum. A Scrum Team may employ this if that helps meeting its Definition of Done and increase the quality. Correct answer is 'b.'

-------------------Question- 7-------------------------

Scrum – Open Questions

There are some questions that are not clarified in Scrum. You may want to be aware of them so you can use Scrum values and high-level principles to infer guidelines around them. That understanding will be handy if there are any related questions in the assessment.

<u>Example 1: In which event are the cost and the number of Sprints decided?</u>

Parameters like release date, cost, etc. are reviewed in the Sprint Review. However, Scrum leaves it open about their first definition. Unless otherwise inferred through the question, it is safe to assume that they are defined at the time when the Just Enough Backlog is defined before the first Sprint.

<u>Example 2: How are defects handled?</u>

- Scrum is a product development framework and not a software development method. Hence 'The Scrum Guide' does not detail the process of how to handle bugs/defects since the definition of bugs/defects are different in different product development sectors.

- When techniques such as Burn-down were changed to optional in 'The Scrum Guide 2013,' the Scrum authors provided the following explanation:

"*It is expected for the Scrum Teams to figure out the best methods, techniques, and practices themselves. While it is optional to apply the technique that works for their team, Scrum does not bind the Scrum users to specific sub-techniques. The reason is Scrum is lightweight and can be implemented even without the need of sub-techniques.*"

There are many sub-techniques of product development like handling defects that are intentionally left undefined. So, by providing a detailed guideline for handling bugs, we tend to add more rules than necessary to keep Scrum lightweight.

Here are some high-level guidelines to answer related questions.

1. Current Sprint defects: These are the defects from the work performed in the current Sprint. The Developers fix them as part of the current Sprint.

The Developers make an informed decision about forecasting what they can accomplish in the current Sprint. There is less likelihood of large defects being introduced because the team has refined the Product Backlog Items to a fine degree so that they can see through the requirements. So, when the team does get defects, it is expected that the defects are minor in nature and the team fixes them in the current Sprint. However, in complex problems, what can happen is unknown. So, if the team does identify major defects, i.e., those that will be considered as new work, they are added to the Product Backlog.

2. Production defects: These are existing production defects of the product. If the defect fixing effort is negligible, the team fixes them in the current Sprint. If it is a time-consuming defect and if completing the current Sprint without addressing this production defect does not makes sense, the Sprint Goal automatically becomes obsolete. The Product Owner is responsible for maximizing the value of the product and the work of the Developers. So, the Product Owner will usually cancel the current Sprint and have the team work on this new business imperative.

3. All other defects: They include the defects from the prior Sprints, non-critical production defects, etc. These are treated as new work to be performed and hence they are added to the Product Backlog.

-------------------Question- 8--------------------------

In the middle of the Sprint, the Developers identified a defect. Initially they were not sure about the cause of the defect, and hence they involved the Product Owner to discuss it. After the discussion, both the Product Owner and the Developers agreed that the defect is indeed a clear gap in the feature being developed in their Sprint and not a new requirement. The defect should be

a) Deferred to the Product Backlog since it is a new requirement.

b) Deferred since this is not a critical defect.

c) Fixed in the current Sprint.

-------Answer-------

This defect is from the work performed in the current Sprint. Correct answer is ‘c.’

-------------------Question- 8--------------------------

Customizing Scrum for special situations

Some teams may have special roles like an impediment manager. Sometimes they create a sub team for production support (Business as usual teams). However, Scrum is immutable and does not allow customization.

-------------------Question- 9--------------------------

A Scrum Team often gets some production support requirements in addition to the work in the Sprint Backlog. The team adapted their team composition and created a sub team to support these ad-hoc requirements.

a) It is okay to create sub teams within a Scrum Team.

b) It is not okay since there cannot be sub teams within a Scrum Team.

c) The team can complete the production support as one team, since it is high priority, and then come back to the original Sprint work.

-------Answer-------

Every Sprint is meant for delivering a valuable and useful product Increment as required in the Sprint Goal. If an outside item is taking the team's time, it is treated as an issue. If the team is forced by any authority, the Scrum Master needs to coach them about how Scrum works and facilitates removing this issue. Correct answer is 'b.'

-------------------Question- 9--------------------------

Iteration/Sprint zero which does not produce a working Increment

Many professionals with on the job Scrum experience may have seen a Sprint called Sprint zero. This Sprint zero is created to accomplish some upfront preparations before the other Sprints. Some of the upfront preparation includes tasks such as setting up the work environment, staffing people, etc. This Sprint zero will not produce a working Increment. This is common in many organizations following Scrum. However, if a Sprint is not intended to create a valuable and useful Increment, Scrum does not acknowledge (call) it as a Sprint.

-------------------Question- 10--------------------------

The architectural features of the product need to be

a) Evolved along with the Sprint deliveries.

b) Completely designed upfront before the Sprints.

c) Decided at least at a skeleton level in Sprint zero.

-------Answer-------

Some teams include an iteration called Sprint zero before the first Sprint to do the initial design. This is the copycat of the traditional "Big Upfront Design" of waterfall. Such practices defeat the purpose of empiricism. Correct answer is 'a.'

-------------------Question- 10--------------------------

Hardening Sprint

Some teams customize Scrum to include an exclusive Sprint for increasing quality called the "Hardening Sprint." The Hardening Sprint focuses on 'perfecting' the Increment to meet the production release requirements. The team takes the Increment that was approved by the Product Owner in the Sprint Review and performs a list of "post Sprint Review" activities to enhance the technical fit of the Increment so it can go to production. However, in Scrum the purpose of every Sprint is to deliver an Increment that meets the quality requirements to be useful. This means that every Increment coming out of any Sprint must be in a valuable and useful state without the need for additional work. The Definition of Done maintained by the Scrum Team should include conditions to meet these quality measures consistent with the organization's quality objectives.

-------------------Question- 11--------------------------

After the Sprint Review, production release in Scrum requires

a) Hardening Sprints.

b) Non-Functional Testing.

c) Architectural Validations.

d) Usability/End User testing.

e) All of the above.

f) None of the above.

-------Answer-------

Every Sprint must produce an Increment that is valuable and useful. This means that the Increment is potentially releasable to production with minimal work. Correct answer is 'f.'

-------------------Question- 11--------------------------

Chapter 4 – Additional Tips

This Chapter is meant to fully prepare you for the assessment. It provides additional material and more assessments to practice.

Changes in The Scrum Guide

The Scrum Guide is continuously updated. The latest version is November 2020.

The older versions are removed and NOT available at scrumguides.org. However, the changes between the versions are documented at https://www.scrumguides.org/revisions.html.

This book is updated with all the changes until the latest version – November 2020. You don't need to know the changes between the versions. Just to be on safer side, a short list of the changed terms between the latest version and the prior versions is given below, so when you see any of these terms in the assessment, whether old or new, you know that you can interpret them as the same.

Term in Latest Version vs. Term in Old Versions

- Developer – was called Development Team Member
- Self-Managing – was called Self-Organizing
- Backlog Refinement – was called Backlog Grooming
- Sprint Forecast - was called Sprint Commitment
- Events - was called Ceremonies
- Definition of Done – was called definition of "Done" (in quotes)

Some of the mandatory elements in the older versions are optional now

The older versions of The Scrum Guide had something called "strategies and tips." They contained the discussion of some techniques and practices that can be employed within the broader framework of Scrum as guidelines. An example of a practice is Release Planning.

The latest version does NOT contain "strategies and tips." It was removed to communicate that Scrum is lightweight and can be implemented without the need for specific techniques provided in "strategies and tips." While it is optional to apply the technique that works for the team, Scrum does not bind the Scrum users to specific sub-techniques. The following list contains items that are now optional:

- Release Planning (Removed)
- Velocity (Made optional)
- Burn-up and Burn-down (Made optional)
- Three questions in Daily Scrum (Made optional)
- At least one improvement item (from Retrospective) in Sprint Backlog (Made optional)

New elements added in the latest version (2020)

- In older versions Sprint Planning was considered to have two separate parts called Part 1 (What) and Part 2 (How). In the 2017 version, Sprint Planning still had these two parts "What and How" as two topics that need to be addressed by the team. But it was not mandatory to separate them into two separate exclusive sessions within Sprint Planning. The team can address the two topics in seamless one session. In the latest version (2020), "Why" (Sprint Goal) is called out explicitly as third topic along with "What and How."

- Product Goal - provides focus for the Scrum Team toward a larger valuable objective. Each Sprint should bring the product closer to the overall Product Goal.

- Commitments - Each of the three artifacts now contain 'commitments' to them. For the Product Backlog it is the Product Goal, the Sprint Backlog has the Sprint Goal, and the Increment has the Definition of Done (now without the quotes). They exist to bring transparency and focus toward the progress of each artifact.

Word Play

For non-native speakers of English, some of the wording may be confusing. Here is the list of some potential phrases and what they mean:

- Increment / Done Product Increment / Done Working Increment / Potentially Shippable / Potentially Releasable – All these phrases mean the same thing which is "valuable and useful product Increment."

- Product Development - Could mean both product development and software development.

- Development Team Member - is same as a Developer. They were interchangeably used in the previous versions of The Scrum Guide.

Please note that if your native language is not English, and if The Scrum Guide is available in your language, read The Scrum Guide first in your native language to grasp some concepts that may otherwise be missed due to language nuances. As of the writing this book, The Scrum Guide is available in several languages other than English.

Some optional standards

- Definition of Ready: Scrum explains that the Product Backlog Items are refined in an act called 'Backlog Refinement' until they are "Ready" for selection in Sprint Planning. There is no explicit reference to anything called "Definition of Ready."

Definition of Ready helps in two ways:

It provides a shared understanding between the Product Owner and the Developers regarding the preferred level of description and transparency each item should meet before introducing them in Sprint Planning. The Team can use this standard as a guideline to refine (decompose) the items.

It provides the transparency to the team so they can estimate the effort and verify that they can get them Done within a Sprint. When a Product Backlog Item reaches this level of transparency, it is also known as "Ready."

- Team Velocity: The Scrum Guide explains that Scrum Teams always try to optimize their productivity, and the Scrum Master causes changes that enhance their productivity. However, it does not mandate any standard to measure the team's productivity. Velocity is an optional standard used within Scrum. Some teams use it to track their development speed and as a basis for their productivity to make a forecast during Sprint Planning. Velocity is usually an absolute number that represents the average quantity of work of the Developers. It shows the average amount of Product Backlog turned into an Increment during a Sprint. The Developers track this to project the capacity of them so they can make a forecast.

• Coding Standards, Architectural Guidelines: The Scrum Guide does not talk about the technical practices or artifacts. There is no reference to anything called "Coding Standards" or "Architectural Guidelines." However, many Scrum Teams use them as part of their technical practices. Coding Standards ensure that the team produces readable and maintainable code. In Scrum, architecture is evolved continuously throughout the product development duration as more is learned. There is no exclusive Sprint or Scrum event to define architecture upfront. Usually, the Scrum Team defines architectural guidelines that every team member can use in their work. The team can have core hours to review the design and architecture during the Sprint. As a result, these guidelines are continually updated.

Scrum for Large efforts (Scaling Scrum)

Though questions are unlikely on this topic on the assessment, there were a few occasions where some PSM 1 assessment takers informed me about one question they face on large scale Scrum. So, it helps to know a little information on this which should help if a question or two appears in the exam. Scrum.org provides a framework called Nexus to scale Scrum for large product development efforts.

At Scrum.org, originally there was a Scrum practitioners Open which later was replaced with the Nexus open assessment. You can also practice the Nexus open assessment available at Scrum.org. It is not a "MUST DO" to prepare for the exam.

Some basic information on structuring the team is given below:

Structuring a large team into feature teams helps minimize or eliminate dependencies. Dividing on other basis such as a technical component is called layer teams or component teams. The advantages of feature teams are

- They are usually self-sufficient and hence have low communication overheads with external teams.
- There will be increased opportunity for direct business collaboration.
- Any dependencies that still exist between feature teams are made transparent and planned in an event called Scrum of Scrums.

-------------------Question- 12--------------------------

Hundreds of developers are identified for a Scrum work. Which two of the following may be appropriate considerations to form these developers into teams?

a) Each team must have a required number of technical leads.
b) Each team must be sized to reduce external dependencies with less internal communication issues.
c) Each team must be a business feature team.
d) The team formation should seek input from the business side.
e) Each team must be a technical component team.

-------Answer-------

There is no technical lead role in Scrum. Hence choice 'a' is incorrect.

Feature team, though a preferred practice, is not a mandatory requirement by Scrum. Hence, choice 'c' is incorrect.

A technical component team increases dependency and reduces the ability for the team to produce a fully integrated working Increment. Hence choice 'e' is incorrect.

Correct answers are 'b' and 'd.'

-------------------Question- 12--------------------------

-------------------Question- 13-------------------------

A large-scale product development requires more than 100 Developers. What is the most appropriate approach to develop an overall technical architecture?

a) Start the product development with the minimal number of teams possible. Let them evolve the foundation architecture that reflects the core product features of high value and commonly expected non-functional needs. Gradually add more teams.

b) Create a complete reference architecture before the development. Provide training to the Developers to teach them to comply with this architecture and hand over the architecture to them.

c) Identify a small set of best designers and let them guide the Developers during the Sprint with its implementation.

d) Divide the teams into technical component teams with specific responsibilities to design and manage their own components. Resolve any ongoing integration issues using Scrum of Scrums.

-------Answer-------

Scrum recognizes no titles for the technical members of the Scrum Team other than 'Developer' regardless of the work being performed by the person. There are no exceptions to this rule. As for the technical architecture, the design emerges throughout the journey The Developers do not create a big upfront design before they start Sprints. Instead they evolve the design.

Given this, choices 'b' and 'c' are incorrect. There is no designer or design team.

Technically dividing the team increases the dependency between the teams in large scale Scrum. Hence choice 'd' is incorrect. Correct answer is 'a.'

-------------------Question- 13-------------------------

-------------------Question- 14-------------------------

When multiple Scrum Teams are working on the same product, how many Product Owners and Product Backlogs are needed?

a) Multiple Product Owners and multiple Product Backlogs.

b) One Product Owner and one Product Backlog.

c) Multiple Product Owners and one Product Backlog.

d) One Product Owner and multiple Product Backlogs.

-------Answer-------

If there is one product, then there must be only one Product Backlog and one Product Owner. All the teams must work from the same Product Backlog. Also, note that it is not necessary for the Definition of Done to be same but a mutually defined Definition of Done should enable the combined Increments to be valuable and useful. Correct answer is 'b.'

-------------------Question- 14-------------------------

-------------------Question- 15-------------------------

When multiple Scrum Teams are working on the same Product Backlog, each team selects the Product Backlog Items for the Sprint with the guidance of the Product Owner.

a) True

b) False

-------Answer-------

The Product Backlog is continuously refined to a thinly sliced functionality so that each Product Backlog Item has very minimal dependency between the Scrum Teams. The refinement also strives to identify which team will deliver what item. Later, in Sprint Planning, each Scrum Team selects the Product Backlog Items with the guidance of the Product Owner. Correct answer is 'a.'

-------------------Question- 15--------------------------

Knowledge of the Agile Manifesto

Scrum mutually respects the Agile Manifesto as a newer way of working. For the questions on the assessment, if the answer is not deductible from within the boundary of Scrum, apply the knowledge from the Agile Manifesto as a second body of knowledge.

Complementing ideas between Agile and Scrum - Four Values

The Agile Manifesto has four values and twelve principles. These are provided below. Scrum implements many of the Agile Manifesto values and principles in terms of Scrum Teams and their associated roles, events, artifacts, and rules.

Agile Value 1: For a better way of building software, the Agile Manifesto values “Individuals and interactions over processes and tools.” Scrum implements this value through Self-Managing teams.

Agile Value 2: For a better way of building software, the Agile Manifesto values “Working Software over comprehensive documentation.” Scrum implements this value through Sprints: A Sprint is a container of a few weeks of development work, where a Done, valuable, and useable product Increment is created. The mark of progress is the creation of this Done Increment, and not the creation of documents.

Agile Value 3: For a better way of building software, the Agile Manifesto values “Customer collaboration over contract negotiation.” Scrum implements this value through the Product Owner role: The Product Owner maximizes the value of the product being developed and optimizes the work of the team through continuous collaboration with the Developers.

Agile Value 4: For a better way of building software, the Agile Manifesto values “Responding to change over following a plan.” Scrum implements this value through Scrum events: Each Scrum event is an opportunity for inspection and plan adjustment. Also, each Scrum event is time-boxed to eliminate unnecessary and extensive planning.

Complementing ideas between Agile and Scrum - Twelve Principles

1. “Our highest priority is to satisfy the customer through early and continuous delivery of valuable software.” Scrum continuously delivers a useable Increment through Sprints which are produced at least once a month.

2. “Welcome changing requirements, even late in development. Agile processes harness change for the customer's competitive advantage.” Scrum absorbs the latest insights and needs into the Product Backlog at anytime.

3. “Deliver working software frequently, from a couple of weeks to a couple of months, with a preference to the shorter timescale.” Scrum continuously delivers a useable Increment through Sprints that are not more than 1 calendar in duration.

4. “Business people and developers must work together daily throughout the project.” Scrum emphasizes the role of Product Owner around maximizing the value of the work by the Developers through Product Backlog management and helping them to understand the business needs.

5. “Build projects around motivated individuals. Give them the environment and support they need, and trust them to get the job done.” Scrum Teams are self-managing where no external authority can direct them on how to perform their work.

6. “The most efficient and effective method of conveying information to and within a development team is face-to-face conversation.” Scrum structures the team such that the communication is less complex and highly effective.

7. “Working software is the primary measure of progress.” Scrum squarely positions the Increment as the only mark of progress.

8. “Agile processes promote sustainable development. The sponsors, developers, and users should be able to maintain a constant pace indefinitely.” Scrum empowers the team with both responsibilities and privileges so that they can make informed and realistic decisions about work without burning themselves out on unrealistic and externally enforced expectations.

9. “Continuous attention to technical excellence and good design enhances agility.” Scrum emphasizes the continuous inspection and adaptation of the Scrum Teams themselves focusing on quality, creativity, and productivity through Sprint Retrospectives.

10. “Simplicity--the art of maximizing the amount of work not done--is essential.” Scrum is intentionally lightweight avoiding thick and heavy processes and tools which have questionable value.

11. “The best architectures, requirements, and designs emerge from self-organizing teams.” Scrum Teams are self-managing teams.

12. “At regular intervals, the team reflects on how to become more effective, then tunes and adjusts its behavior accordingly.” Scrum Teams perform a mandatory Sprint Retrospective every Sprint which is an inspection and adaptation of the team itself.

Additional knowledge from the “Developer Open”

The PSM 1 may contain one or two questions from the Scrum development work perspective. If you have time, attend the Developer Open assessment to grasp some of the related concepts.

#

Chapter 5 – Practicing Quick Tests

Here are five short tests to test your Scrum knowledge. If you score high in the first test, i.e., a minimum of 8, then you can go to next quick test. If you score less than eight in a quick test, analyze what you missed. You can go back to the relevant sections in the previous chapters to reinforce your learning, and come back to the quick tests again. Finally, you can take one full-blown model assessment that resembles the real assessment. If you get through that with a score of 68+ correct questions, you are well prepared.

#

Chapter 5.1 - Quick Test 1

Quick Test 1 - Questions

1. The role of Scrum Master in the Sprint Retrospective is

a) Auditor.

b) Silent Observer.

c) Peer Team Member.

d) None of the above.

2. To deliver a single product, three different Scrum Teams are formed. How many Product Owners are needed?

a) As many as recommended by the Scrum Master.

b) Three.

c) One.

3. Scrum framework is founded on

a) Empiricism.

b) Empiricism and Technical Practices.

c) Empiricism and Emotional Intelligence.

4. After Sprint Review, Production release in Scrum requires

a) Hardening Sprints.

b) Non-Functional Testing.

c) Architectural Validations.

d) Usability/End User testing.

e) All of the above.

f) None of the above.

5. A Scrum Team crafts the following Sprint Goal: "All the Sprint code should have passed 100% automated unit tests."

a) Not an appropriate goal since the Sprint Goal should be about expected business value.

b) It is incorrect since the Product Owner formulates the goal and not the Scrum Team.

c) It is a well-formed Sprint Goal.

6. One of the Scrum Teams chose to have a Developer also play the role of Scrum Master part time. A Developer cannot also play Scrum Master's role.

a) True.

b) False.

7. The duration (length) of the Sprint is decided by the

a) Product Owner.

b) Scrum Master.

c) Scrum Team.

8. During the Daily Scrum, which plan is used as a reference to understand the changes in progress?

a) Sprint Backlog

b) Product Backlog

c) Sprint Burn-down

9. An important executive wants the Developers to take in a highly critical feature in the current Sprint. The Developers

a) Should work on that since the organization's priority is more important.

b) Should ask the executive to work with the Product Owner.

c) As an empowered team they should ask the executive to select work to be removed so that there will be the same amount of work in the Sprint.

10. A Scrum Team is in the process of defining Product Backlog Items. The Scrum Master notices that the team is not using a User Story format to capture the backlog items. The Scrum Master should

a) Correct the team's behavior by coaching them about User Stories.

b) Let the team decide the format of Product Backlog Items.

c) Add a business analyst with knowledge of writing User Stories to the team with a specific responsibility of documenting backlog items in terms of user stories.

Quick Test 1 - Answers

1. Correct answer is 'c.' One of the items reviewed in the Sprint Retrospective is the "implementation of Scrum framework." Since the Scrum Master is the owner of that, they participate as a peer team member

2. Correct answer is 'c.' A single product should have a single Product Backlog and hence only one owner, a Product Owner. A Product Owner can delegate some of their responsibilities to the team, however they are still accountable for the Product Backlog ownership.

3. Correct answer is 'a.' Technical practices or any other value adding techniques can be optionally chosen by the Team and followed within the Scrum framework. However, they do not reflect the foundation of Scrum.

4. Correct answer is 'f.' Every Sprint produces a valuable and useful Increment.

5. Correct answer is 'a.' The Sprint Goal reflects the intended business functionality that will be delivered in a Sprint.

6. Correct answer is 'b.' A Scrum Master can be a Developer but that is not mandatory.

7. Correct answer is 'c.'

8. Correct answer is 'a.' The Sprint Backlog is a plan with enough detail that changes in progress can be understood in the Daily Scrum.

9. Correct answer is 'b.'

10. Correct answer is 'b.' Scrum does not prescribe any specific technique to capture the Product Backlog Items. The team can choose the most beneficial technique that works for them.

#

Chapter 5.2 - Quick Test 2

Quick Test 2 - Questions

1. The Scrum Team uses the information of Scrum artifacts to make ongoing decisions. The soundness of these decisions depends on the

a) Artifacts' Adaptability.

b) Artifacts' Transparency.

c) Artifacts' Agility.

d) Artifacts' Format.

2. An organization decides to have small Scrum Teams with less than three Developers. The likely result could be

a) The team may have decreased interaction.

b) The team may have a skills shortage.

c) The team may have low productivity gains.

d) All of the above.

3. The product development project is about delivering an internal feature for an organization. The team has good skill composition and worked on similar projects. The Sprint lengths can be

a) Decided after the first release.

b) Ignored since it is an internal project.

c) A maximum of one calendar month.

4. Select all that apply. Empiricism provides

a) Frequent opportunities to obtain information so uncertainty can be completely eliminated.

b) Frequent opportunities to discuss different possibilities.

c) Frequent opportunities to make informed decisions reducing risk.

5. The leadership model followed by Scrum Master is

a) Micro-management.

b) Servant Leadership.

c) Command and Control.

6. During a Sprint Review, the stakeholders notice that the product development progress is not clearly visible and lacks transparency. Moreover, they are not able to understand the next steps. Who is responsible for this?

a) Developers

b) Product Owner

c) Scrum Master

d) Scrum Team

7. In the middle of the Sprint, a team member was required by another department manager to support an important task

outside the Sprint work. What is recommended for the team member to do?

a) The team member must support since it is important task.

b) The team member should ask the manager to speak with the Scrum Master.

c) The team member should politely decline and explain to the manager his responsibility and accountability to the Scrum Team.

8. When more Scrum Teams are added to a project that works on one single product, the productivity of the original Scrum Teams mostly likely will increase.

a) True

b) False

9. Within just a few Sprints, Scrum increases the transparency of the following

a) Technical ability of team to create a Product Increment.

b) Information of real progress.

c) Both.

10. The architectural features of the product need to be

a) Evolved along with Sprint deliveries.

b) Completely designed upfront before the Sprints.

c) Decided at least at a skeleton level in Sprint zero.

Quick Test 2 - Answers

1. Correct answer is 'b.' Significant aspects of the development process must be visible to those responsible for the outcome. These aspects must be highly transparent so appropriate decisions can be taken.

2. Correct answer is 'd.' While the Scrum Team should be small enough to be nimble, fewer than three Developers decreases interaction and results in smaller productivity gains. Smaller Scrum Teams may encounter skill constraints during the Sprint causing the Developers to be unable to deliver a valuable and useful Increment.

3. Correct answer is 'c.' Sprint length should be decided for all Sprints including for the first Sprint. Sprints are limited to one calendar month. The Product Owner's input is needed to verify that the business risk due to this Sprint length is acceptable. In this case, since the team is cross-functional and experienced, risk appears to be lower. So, the length can be shorter.

4. Correct answers are 'b' and 'c.' Empiricism is an alternative to waterfall to manage complexity and uncertainty. In waterfall, the risk of uncertainty accumulates over long cycles. The risk is reduced by providing frequent feedback and course correction points where more information may be available to view different possibilities and make informed decisions. However, empiricism does not completely eliminate uncertainty.

5. Correct answer is 'b.' The Scrum Master is a servant-leader for the Scrum Team.

6. Correct answer is 'b.' The Product Owner is responsible for maintaining the transparency of the Product Backlog, the progress so far, and the next steps along with alternatives if any.

7. Correct answer is 'c.' Other than the "Backlog Refinement" the Developers should only work on the tasks related to the Sprint Goal. If any external authority brings different work, the team should explain that they self-manage their work in a Scrum way. The team can refer them to the Product Owner to talk about adding the work to the Product Backlog.

8. Correct answer is 'b.' Each Scrum Team needs to mutually define their Definition of Done so their combined work will be valuable and useful. This involves some overhead work in syncing up and may impact productivity.

9. Correct answer is 'c.' Since a Sprint completes one full cycle of development activities including Sprint Planning, developing, delivering a valuable and useful Increment, etc. useful information and transparency is available.

10. Correct answer is 'a.' Some teams may customize Scrum to include an iteration called Sprint zero before the first Sprint to do design work. This is a replacement of the traditional "Big Upfront Design" of waterfall and defeats the purpose of empiricism.

#

Chapter 5.3 - Quick Test 3

Quick Test 3 - Questions

1. A Sprint that is longer than one calendar month may result in

a) Too much to inspect in short meetings.

b) Detached stakeholders.

c) Increased complexity needing more traditional controls like documentation.

d) All of the above.

2. The work left against time is shown by

a) Team Velocity.

b) Burn-down graph.

c) Story Points Burn.

d) Release Burn-up.

3. In the Sprint Review, along with the review of the product Increment and progress, "what (steps) to do next" is also discussed.

a) False

b) True, and the scope of the next Sprint is also finalized here.

c) True, and it may capture probable Backlog Items for the next Sprint, but the scope of the next Sprint is deferred until Sprint Planning.

4. In the middle of a Sprint, the Product Owner wants the Developers to participate in an important meeting with a group of customers.

a) The Developers should strive to work on items related to current Sprint Goal. They should involve the Scrum Master who can educate the Product Owner to defer such interruptions and, if required, plan them in the next Sprint.

b) The Developers should participate in the meeting since it is with customers.

c) The Developers should stop the current Sprint work until that meeting to ensure they clearly understand the customer's concerns.

5. A Scrum Team decides that the frequency of the Daily Scrum should be reduced to once a week.

a) The Scrum Team is self-managing and can choose their own practices.

b) Self-management is alright, but such decisions need to be approved by the Agile Coach. So, they should involve the Agile Coach.

c) Self-management is about how to get the Sprint work done but must follow Scrum. So, the Scrum Master should strive to coach the team on the essentials of the Daily Scrum.

6. Who performs inspections of the work in Scrum?

a) External Audit Team

b) Scrum Master at defined inspection points

c) Developers

7. During the Daily Scrum, a team member says he does not know when his task will be complete.

a) It is acceptable as the Sprint Review date is far away.

b) Replace the team member with a new team member.

c) The Developers should collaborate to plan alternative steps such as pairing this member with someone else, etc. to eliminate the risk of not meeting the Sprint Goal.

d) Ask the Scrum Master to mentor the team member on how to estimate the task.

8. In every Sprint, the working Increment should be tested progressively from unit testing, to integration testing, and then user acceptance testing.

a) Yes. It is the prescribed method.

b) No. The test strategy is decided by the Quality Assurance Lead.

c) Not necessary. While the team needs to ensure that each Increment is thoroughly tested, all Increments work together, and meets the Definition of Done, it is up to the team to find best method to achieve this.

d) Incorrect. It should also include non-functional testing.

9. You are on a Scrum Team that is in the middle of a Sprint. Your team gets some additional team members. The likely result is:

a) The team can take more stories on top of the originally forecast Sprint Backlog.

b) The team will have to do Sprint Planning again and get buy-in from the new members on the new planning.

c) The team may suffer in its productivity.

10. A Scrum Team needs to develop a web application in Increments. Some of the Sprints have Sprint Goals like: 'Develop Data layer for Functionality A'. What is your inference?

a) The Scrum Team follows horizontal decomposition of the Product Backlog Items. This is recommended.

b) The Scrum Team follows vertical decomposition of the Product Backlog Items. This is recommended.

c) The Scrum Team follows horizontal decomposition of the Product Backlog Items. This is NOT recommended.

d) The Scrum Team follows vertical decomposition of the Product Backlog Items. This is NOT recommended.

Quick Test 3 - Answers

1. Correct answer is 'd.' The longer the Sprint length, the work practices tend to drift towards waterfall style: with lengthy meetings, lack of early feedback from stakeholders, documentation/communication needs due to increasing complexity.

2. Correct answer is 'b.'

3. Correct answer is 'c.' Each Sprint event is an opportunity to inspect and adapt. "What to do next" is about adapting the Product Backlog if needed. The scope of the Sprint is finalized in the Sprint Planning and not in the Sprint Review.

4. Correct answer is 'a.' Other than the act of "Backlog Refinement," each task that the Developers perform must be related to the Sprint Goal. Any distraction should be avoided and, if necessary, the Scrum Master's help needs to be sought to educate those causing the disruption.

5. Correct answer is 'c.'

6. Correct answer is 'c.' The Developers are responsible for inspecting their own work.

7. Correct answer is 'c.' The highest priority of the Developers is to complete the Sprint Goal. If there are impediments, they need to be resolved either directly or by asking the Scrum Master for assistance. Later, in the Retrospective, the cause of this impediment can be discussed to find potential improvements.

8. Correct answer is 'c.' The team self-manages its own work. They can employ approaches and techniques that provide the best return on effort.

9. Correct answer is 'c.' The productivity will not increase because there will be a learning curve for the new members. The Sprint cannot be aborted to go back to Sprint Planning. The Sprint can be cancelled only by the Product Owner upon their inference that the Sprint Goal is no longer valid.

10. Correct answer is 'c.' It is preferable to decompose the Product Backlog Items such that each team can produce useable business functionality instead of producing a technical component. Such decomposition based on useable business functionality is called vertical decomposition. A horizontal decomposition, on the other hand, makes the team a technical component team that will have external dependencies.

#

Chapter 5.4 - Quick Test 4

Quick Test 4 – Questions

1. The Definition of Done is
   a) Initially defined per product by the Scrum Team, but may change throughout the product development duration.
   b) Initially defined by the Scrum Team and does not change.
   c) Defined after the first Sprint based on the new insights obtained from the first Sprint Review.

2. Which of the following statements is true? Select all that apply.
   a) After Sprint Planning, a Sprint cannot proceed without complete requirement specification.
   b) After Sprint Planning, a Sprint cannot proceed without a Sprint Goal.
   c) After Sprint Planning, a Sprint can proceed without a complete Sprint Backlog.
   d) After Sprint Planning, a Sprint cannot proceed without complete architecture.

3. A Scrum Team is self-managing and empowered. It means it has the authority on internally deciding who does what, when, and how.
   a) True
   b) False

4. Who decides the duration of the Sprint?
   a) Product Manager
   b) Scrum Master
   c) Developers
   d) Scrum Team

5. A Product Owner is not available for Scrum events and not supportive enough for the Developers. The next immediate accountability is with the
   a) Developers who need to cancel the Sprint.
   b) Stakeholders who need to get a written commitment from the Product Owner.
   c) Product Owner's manager who needs to engage the Developers and understand their problems.
   d) Scrum Master who needs to educate the Product Owner on his role.

6. When is a Product Backlog retired?
   a) When the Product Owner retires.
   b) When all the Sprints are over.
   c) When the Product retires.
   d) When the Customer provides the sign-off on completion of the project.

7. A Product Owner cannot send a representative (delegate) to the Sprint Review.

a) True

b) False

8. A Product Owner is also knowledgeable on technology. In addition to product requirements, they also impose some technical conditions that the product should meet. These conditions must be added to the

a) Product Backlog.

b) Sprint Backlog.

c) Definition of Done.

9. An Increment is

a) The sum of the value of all Increments from previous Sprints integrated with the Done Product Backlog Items in the latest Sprint.

b) The sum of the Product Backlog Items selected into the Sprint Backlog.

c) The sum of the Product Backlog Items Done in the latest Sprint.

10. Which are true statements? Select all that apply.

a) The Scrum Team is responsible for formulating a Sprint Goal.

b) When existing Product Backlog Items in the Sprint Backlog are modified, the Sprint Goal is bound to become invalid.

c) The coherence between Product Backlog Items is made transparent by the Sprint Goal. Lack of coherence will lead to the Developers working individually.

Quick Test 4 – Answers

1. Correct answer is 'a.'

2. Correct answers are 'b' and 'c.'

3. Correct answer is 'a.'

4. Correct answer is 'd.' The final Sprint duration, i.e., how much shorter than one month, is decided by the Scrum Team after considering the need of the Product Owner to limit business risks and the need of the Developers to synchronize the development work with other business events.

5. Correct answer is 'd.' The Scrum Master has the responsibility to remove the Scrum Team's impediments and coach all the Scrum Team members. Also, the Scrum Master can show the poor results, which were due to the lack of Product Ownership, to the Product Owner during the Retrospective.

6. Correct answer is 'c.' A Product Backlog is a living artifact that lives as long the product lives.

7. Correct answer is 'a.' A Product Owner, though accountable for the Product Backlog, can delegate many of the activities around the Product Backlog Management, such writing them, ordering them, etc. However, they cannot delegate their participation in Scrum events.

8. Correct answer is 'c.' Every Product Backlog Item should be about the product's need to bring business value. The condition that the Product Owner brings here is about the technical constraint. So, it should be added to the Definition of Done.

9. Correct answer is 'a.' The Increment is the sum of all the Product Backlog Items completed during a Sprint and the value of the Increments of all previous Sprints. An increment is a body of inspectable, done work that supports empiricism at the end of the Sprint.

10. Correct answers are 'a' and 'c.' The Sprint Goal provides an opportunity for the team members to work together and offers some flexibility of adjusting the Product Backlog Items when required. The Developers can modify the Product Backlog Items in the Sprint Backlog with the Product Owner's consent, but the Sprint Goal must still be met.

#

Chapter 5.5 - Quick Test 5

Quick Test 5 – Questions

1. Which is not a Product Backlog Management activity?

 a) Clearly expressing and ordering Product Backlog Items.

 b) Optimizing the value of the work of the Developers.

 c) Using formal change control to manage Product Backlog when the market provides feedback from product usage.

 d) Ensuring the Developers understand items in the Product Backlog to the level needed.

2. Select all that apply. The Scrum Team must participate in the

 a) Sprint Planning.

 b) Daily Scrum.

 c) Sprint Review.

 d) Sprint Retrospective.

3. An inspector finds that a work aspect deviates outside acceptable limits, and that the resulting product will be unacceptable. When will the team adjust this work aspect to minimize the deviation?

 a) In the next Scrum event.

 b) As soon as possible.

 c) After the Scrum Master approves the adjustment.

4. A Scrum Team can identify the improvements only during the Sprint Retrospective.

 a) True

 b) False

5. For the first Sprint, the inputs are the Product Backlog and the Projected Capacity of the Developers. What are the additional inputs to the subsequent Sprints? Select all that apply.

 a) The defect list from the previous Sprint.

 b) The Sprint Plan.

 c) The past performance of the Developers.

 d) The latest Product Increment.

6. When a Sprint is cancelled, the Scrum Team discards all the work and refines a new Product Backlog.

 a) True

 b) False

7. At the end of Sprint Planning, the Developers could not decompose all the work into units of one day or less. It could decompose the work for only the first few days of the Sprint.

 a) The Developers should close the Sprint Planning and start the work.

 b) Since the team is self-managing, they should continue Sprint Planning in the following days before they start the work.

 c) The Scrum Master should coach the team in required skills.

8. What is a key inspect and adapt meeting for the Developers?
 a) Project Status Meeting
 b) Daily Scrum
 c) Design Sessions

9. Which are true statements? Select all that apply.
 a) Only the Product Owner should update the Product Backlog without delegating to anyone.
 b) Only the Developers should be responsible for estimates of the Product Backlog Items.
 c) Only the Product Owner should cancel the Sprint. Others can influence the decision to cancel.
 d) Only the Product Owner can change the Sprint Backlog.

10. Who defines the Definition of Done?
 a) Developers
 b) Technical/Domain Experts
 c) Product Owner
 d) Scrum Team

Quick Test 5 – Answers

1. Correct answer is 'c.' Changes in business requirements, market conditions, or technology may cause changes in the Product Backlog. The Product Owner keeps the Product Backlog updated as a living artifact to reflect these changes without a formal change control process.

2. Correct answers are 'a,' 'c,' and 'd.' The Scrum Team participates in all events except the Daily Scrum. The Developers must participate in Daily Scrum event because it organizes, plans, and controls its work without the direction or management of the Product Owner or Scrum Master. The Scrum Master can participate if there is a need to coach or facilitate until the Developers can do that on its own.

3. Correct answer is 'b.' The Developers do not wait for any formal event to make this adjustment; instead they make it as soon as possible to minimize further deviation.

4. Correct answer is 'b.' The Sprint Retrospective provides a formal opportunity to focus on inspection and adaptation. However, improvements may be identified and implemented at any time.

5. Correct answers are 'c' and 'd.'

6. Correct answer is 'b.' The team still conducts the Sprint Review to review Done Product Backlog Items. If part of the work is valuable and useful, the Product Owner typically accepts it. All incomplete Product Backlog Items are re-estimated and put back in the Product Backlog.

7. Correct answer is 'a.' Sprint Planning is time-boxed and cannot be extended. It is enough to have the work decomposed for the first few days of the Sprint to start the work and can be decomposed as needed throughout the Sprint.

8. Correct answer is 'b.' Daily Scrums improve communication, eliminate other meetings, identify impediments to development for removal, highlight and promote quick decision-making, and improve the Developers' level of knowledge. This is a key inspect and adapt meeting for the Developers.

9. Correct answers are 'b' and 'c.'

10. Correct answer is 'a.' The Definition of Done is developed by the Developers with conditions that are acceptable to the Product Owner.

#

Chapter 6 – Model Assessment

Model Assessment - Questions

1. The standard used by the Product Owner and the Scrum Team to identify unfinished work in a Sprint is

a) Coding Standard

b) Definition of Ready

c) Testing Standard

d) Definition of Done

2. Scrum is immutable. What may be the result of an organization modifying Scrum Framework in its implementation for the convenience of existing culture?

a) The organization may lose the opportunity to expose its current cultural dysfunctions that impede the ability to develop the Product Increment Sprint after Sprint.

b) Scrum is bound by technical tools and these tools will break

c) It can only be done with the help of Scrum coaches

3. In a Scrum based software project, "Earned Value" is a good metric to track product development progress

a) Yes

b) No

4. The Scrum Master manages

a) Scrum People

b) Scrum Framework

c) Scrum Technology

d) All of them

e) None of them

5. Select all that apply. Which Scrum events facilitate inspection and adaptation?

a) Sprint

b) Backlog Refinement

c) Sprint Retrospective

d) Development Work

6. The Sprint Review is an opportunity to review

a) Timeline and Budget

b) Defects and causes

c) Requirements and Capacity

d) All of the above

7. The Scrum Team optimizes the following and deliver business value

a) Flexibility, creativity, and productivity

b) Self-Improvement, Leadership, Motivation

c) Individual Power, Heroic Efforts, Recognition

8. Scrum allows having gaps between two subsequent Sprints, in which the team can accomplish support activities and team building activities

a) True

b) False

9. Sprint Planning helps in

a) Building entire technical architecture

b) Staffing plan

c) Testing strategy

d) Release plan

e) None of the above

10. When can a Product Owner negotiate the scope of what the team will work on next?

a) Anytime during the current Sprint with or without Developers' consent

b) Until the Sprint Planning for the current Sprint

c) Both

11. The Developers have not completed any of the Product Backlog Items selected for the Sprint by Sprint end. Next step is

a) Extend the Sprint since Scrum favors "getting done"

b) Advice the Product Owner to accept the completed portion of the incomplete Product Backlog Items, and plan to complete them by next Sprint, since Scrum favors "empowered teams"

c) End the Sprint with a Retrospective, since Scrum favors "time boxing"

12. The Scrum Team, based on the learning from previous Sprints, decides to revisit the length of the Sprint. What is the appropriate Scrum event to discuss and agree on the change?

a) Scrum Planning

b) Sprint Planning

c) Retrospective

d) Daily Scrum

13. To effectively track the Sprint progress, Scrum mandates

a) Preparing Sprint burn down charts

b) Increasing the transparency by frequently updating the remaining work

c) Earned Value approach

14. Only the Product Owner can come up with items that can be considered for Product Backlog. Others cannot provide input / recommendations / ideas about new items

a) True

b) False

15. Sprint Planning is the only occasion where the Developers estimate the Product Backlog Items

a) True, because without estimate, the team cannot plan what can go into the Sprint

b) False, estimation of Product Backlog Items is a continuous event throughout

16. Which is true?

a) Sprint Retrospective focuses on Product and Sprint Review focuses on development processes

b) Sprint Retrospective focuses on development processes and Sprint Review focuses on Velocity

c) Sprint Retrospective focuses on development processes and Sprint Review focuses on Product

17. A Scrum Team often runs into following issues: Conflicting requirements from different departments, ad-hoc work requests from different business managers, no feedback on Increments. What could be the likely cause?

a) Issues with how Scrum Master guides the team

b) Issues with Product Owner responsibilities

c) Issues with planning abilities of the Developers

18. During a Sprint Review, the Scrum Master notices that the Product Owner does not use the Product burn-down graph to explain the status to the stakeholders. The Scrum Master

a) Should coach the Product Owner on the importance of using this Scrum tool

b) Should cancel the Sprint Review and schedule it back when the Product Owner is ready with this tool

c) Do Nothing

19. A short expression of the purpose of a Sprint which is often a business need– is called

a) Sprint Goal

b) Acceptance Criteria

c) Definition of Done

20. The estimation method recommended by Scrum is

a) Planning Poker

b) T-Shirt Sizing

c) Yesterday's weather

d) None of the above

21. It is mandatory that the Definition of Done includes "Release to Production"

a) Yes

b) No

22. Under this topic of the Sprint Planning, the Developers are more active in planning and Product Owner is mostly observing or clarifying

a) Topic One (Why)

b) Topic Two (What)

c) Topic Three (How)

23. Definition of Done is

a) Testing strategy for Scrum Team

b) A standard used by Scrum Team to assess if a product Increment is Done

c) Defined by Product Owner and safeguarded by Scrum Master

24. Shortly into using Scrum for the first time in an organization, the Scrum Team runs into several impediments in following Scrum. The most common inference is

a) Scrum does not work for their organization

b) The Scrum Team didn't plan the project end-to-end well in advance

c) It is normal for first timers. Scrum will expose all weakness in the current eco-system that impede developing Product Increments in short Sprints.

25. A person external to the Scrum Team with a specific interest in and knowledge of a product that is required for Incremental discovery, is known as

a) Technical/Domain Expert

b) Stakeholder

c) Senior Management

26. On their kick-off day, a new Scrum Team didn't have any Scrum tool. The next best thing to do is

a) Expedite the installation of tool before the close of iteration zero

b) Get the recommendation from Product Owner about how to manage Scrum artifacts without the tool

c) Do nothing. Implementation of Scrum does not require any tool

27. The Developers try to put together some guidelines on testing approach. Who will own these guidelines?

a) Developers

b) Test Lead

c) Scrum Master

28. Select all that apply. The mandatory participants of the Sprint Retrospective meeting are

a) Product Owner

b) Stakeholders invited by Product Owner

c) Scrum Master

d) Developers

e) Technical/Domain/Process experts invited by Scrum Team

29. Sprint Backlog is modified throughout the Sprint. As soon as a new task is identified,

a) Product Owner adds it to the Sprint Backlog and communicates about it to Scrum Team

b) Scrum Master adds it to the Sprint Backlog and communicates about it to Scrum Team

c) Developers add it to the Sprint Backlog and communicate about it to Scrum Team

30. Select all that apply. The Sprint Review is an event that requires

a) Product Owner's sign-off

b) Stakeholders active participation

c) Transition sign-off

d) Inspection and Adaptation activities

31. Multiple Scrum Teams are required to work on the same product. How can they integrate their development?

a) work with each other to create an integrated Increment

b) maintain individual Product Backlog for each team

c) set up some working sessions between the lead Developers of each team to merge their changes before the Sprint Review

32. The Sprint Backlog emerges during the Sprint because the Developers modify it throughout the Sprint. In the middle of the Sprint, new work is added to Sprint Backlog. As a result, estimated remaining work will

a) Increase

b) Decrease

c) Stay the same

33. A Scrum Team develops software. Only when the Product Owner decides to go for the release, the team creates end user documentation for the Product Increment at that point.

a) It is correct. Creating document early will require constant effort to keep them updated.

b) It is correct. Scrum favors less documentation and deferring the decision to last minute.

c) It is incorrect. Anything required for the Product Increment to be valuable and useful (potentially releasable to production) is recommended to be part of Definition of Done.

34. Pick the Scrum Values

a) Respect and courage

b) Simplicity

c) Commitment and Openness

d) Creativity and Intuition

e) Focus

35. A Scrum Team has five members. Each one works on a different product. What could we infer about the team?

a) The team will have higher productivity since division of work is clear

b) The team implements diversity, a principle of Scrum

c) The potential of teamwork and benefit of Scrum is less

d) All of them still will have common Definition of Done

36. Team Velocity refers to

a) Average of amount of Product Backlog Items turned into Done Items per Sprint

b) Average rate of churn of team members in Scrum Team during a Sprint

c) Average number of defects per Sprint normalized over all defect types

37. One of the major challenges for the team getting newly into Scrum can be

a) Developing skills to produce useable Increment just within a short Sprint

b) Learning about Scrum terminology

c) Difficulty in getting adapted to Scrum tools

38. In the middle of the Sprint, the Developers find that few more days of work is needed to complete the scope. The planning options include:

a) Add more team members

b) Catch up using weekends

c) Defer the activities like testing after stakeholder's demo

d) Involve the Product Owner and negotiate alternatives

e) All of the above

39. Scrum Master forecasts the Product burn-down during Sprint Review.

a) True

b) False

40. In the middle of the Sprint, the Developers did not get some technical tools that were originally promised. This will slow down the work. The next best thing to do is

a) Scrum Master should escalate to Project Manager

b) Product Owner should cancel the Sprint

c) The Developers should assess the impact to meeting the Sprint Goal and the Definition of Done, and find alternatives to still meet the Sprint Goal without compromising the Definition of Done

41. The Developers of a Scrum Team have created the Sprint Backlog in the form of a task board. What is your inference?

a) The team can choose to represent it any form that makes sense

b) It is okay to have it in task board format, but it must be ensured that it follows Kanban guidelines

c) Scrum Master must coach the team to create proper Sprint Backlog in the form of list of backlog items, related tasks, and estimations

42. The selection of items from the Product Backlog that the Developers deem feasible for implementation in a Sprint is called

a) Estimation

b) Planning Poker

c) Forecast of functionality

43. Velocity is an indication of team performance. It may be used by

a) The Scrum Team an internal measure to plan and track their improvements.

b) The managers to do performance appraisals for the team

c) The organization to aggregate into organization level productivity

44. In a new Scrum Team, a Scrum Master notices that a Developer works on a task that is not contributing to the Sprint Goal or the Sprint Backlog. The Scrum Master

a) Should escalate this to Product Owner

b) Should discuss with team member and educate about Scrum way of working

c) Should not interrupt since the team is self-managing

45. The Developers of a Scrum Team often get some production support requirements, in addition to the work in the Sprint Backlog. The team adapted their team composition and created an exclusive sub team to support these ad-hoc requirements.

a) It is okay to create sub team within Scrum

b) It is not okay since there cannot be sub teams within the Scrum Team.

c) The team can complete the production support as one team, since it is high priority, and then come back to original Sprint work

d) It is okay if it is explicitly approved by Scrum Master

46. A Scrum Team has following condition under the Definition of Done: “All the code to be reviewed and approved by Industry Coding Standard Organization.” This Industry Coding Standard Organization is a third-party Subject Matter Expert outside Scrum Team.

a) The Definition of Done is less effective, because it contains conditions that is not completely within influence of the Scrum Team

b) The Definition of Done is more effective, because it ensures that required standards are met

c) The Definition of Done can contain anything as decided by Product Owner

47. During Sprints, the Developers of a Scrum Team need to wait for another team to provide some dependent input. Often this leads to delay in completing their work. What can be recommended to this team?

a) The team is not cross functional enough. The team should take Scrum Master's help in educating the organization to add team members with appropriate skills

b) The team should agree on Service Level Agreement (SLA) with another team and escalate to Scrum Master if the SLA breached

c) The team can mock up the sample of input instead of waiting and do the Sprint Review on time. The Product Increment can be refactored as and when another team provides input.

48. The Scrum Team gathers for Sprint Planning meeting. The Product Owner has some Product Backlog Items but the Developers find that they do not provide enough information to understand the work involved to make forecast. The next best thing to do is

a) The Scrum Master cancels the Sprint

b) The Developers proceed with starting with whatever is known

c) The Developers make it transparent that they cannot make a forecast with insufficient information, and negotiate with Product Owner on refining the Product Backlog Items to ready state

d) The Scrum Team discusses the root cause in the Retrospective

49. In the middle of the Sprint, the Developers find that they have more capacity to take more work. The next best thing to do is

a) Make it transparent to Product Owner immediately and collaborate to add additional work.

b) Consult and follow Scrum Master's and follow their direction

c) Keep that as a contingency to accommodate unplanned work

50. The Developers are not having regular (Daily) Scrums. As a Scrum Master, you

a) Will advise them to think about conducting regular Scrums, but will let them take the decision themselves as they are self-managing

b) Will escalate this to resource managers

c) Will step in directly to guard the Scrum Framework by asking action-begetting questions to team and positively influencing them to conduct Scrum events

51. When a Scrum Team adds new team members for replacing some members going out, the productivity of the team

a) Will be negatively impacted

b) Will be positively impacted

c) Will remain the same

52. Effort required to fix/refactor something after it has been built is known as

a) Maintenance

b) Technical Debt

c) Plumbing code

53. The role of Scrum Master with respect to Scrum artifacts is

a) Coach the team to increase the transparency of the artifacts

b) Decide the format of the artifacts and ensures that the team follows it

c) Owner of the artifacts and responsible for having them up to date

54. Scrum framework is used to optimize value and control risk in complex product development. A component of value optimization is

a) Averaging out the values delivered over Sprints and use it to take decisions

b) Deciding to continue a Sprint only after verifying if it has enough value worth the effort

c) Ensuring that the Developers are not having idle time by constantly monitoring their productivity

55. Three Scrum Teams are working as part of a big project to develop a product. When Sprints are in motion, there will be

a) Three Product Backlogs, and three Sprint Backlogs

b) One Product Backlog, and three Sprint Backlogs

c) One Product Backlog and one Sprint Backlog

56. Usually, when Scrum is applied newly in an organization,

a) Power of empiricism will be transparent

b) Everything that impedes producing value in short Sprints and accumulation of waste will be made transparent

c) The organization change management process defined by Scrum should be followed to avoid implementation issues

57. In empiricism, the decisions are based on

a) Scientific calculation and Prediction

b) Meeting and Brainstorming

c) Observation, experience, and experimentation

58. What is the correct statement?

a) The technical design continuously evolves over the Sprints. Hence the team should have some basic guidelines to start with, but try to emerge the design through the Sprints.

b) The team can choose to have an exclusive Sprint only to finalize the technical design. At the end, the design should be approved by the project architect

c) The team does not need to pay attention on the architecture as it will evolve itself as a by-product of self-management

59. The Developers of a Scrum Team are often interrupted in the Sprint midway and assigned to work on "other" high priority items. Frequently, such interruptions lead to not meeting the Sprint Goal. The most likely cause could be

a) The Developers are not technically competent

b) The Product Owner authority is ineffective or influenced by another authority

c) The Sprint Planning is poor

60. Select all that apply. The Developers are accountable for

a) Selecting the Product Backlog Items for the Sprint after clarifying with the Product Owner

b) Reporting to the Scrum Master

c) Creating a valuable and useful Increment every Sprint

d) Increasing the productivity as per management goal

61. The process of the coming into existence or prominence of new facts or new knowledge, or knowledge of a fact becoming visible unexpectedly, is called as

a) Transparency

b) Inspection

c) Emergence

62. Middle of the Scrum, the team comes to know that there are some usage related changes to the Product needs. The Product Backlog

a) Is modified to reflect the new need

b) Is closed. Project is cancelled and new Product Backlog will be built

c) Is not impacted and the Sprints continued

63. Middle of the Sprint, the Developers find that some of the Product Backlog Items forecast for this Sprint cannot be finished because they need significant additional effort. However, they can still meet Sprint Goal with rest of the items. The next thing to do is

a) Consult with Product Owner and if they agree, have them cancel the current Sprint, and plan new Sprint with new estimations

b) Do not cancel or modify the Sprint. Extend the Sprint duration as required for the additional effort

c) Collaborate with the Product Owner to remove the Product Backlog Items that cannot progress, and new work up to team's capacity. Complete the Sprint.

64. A good guideline to differentiate Acceptance Criteria from Definition of Done is- Definition of Done provides checklist of quality measures to take the Increment close to usable state, while Acceptance Criteria specify the business functionality

a) True

b) False

65. What is the desirable team composition for large product development program?

a) Program is divided into individual Scrums based on business feature. Each Scrum Team has all the skills needed to finish job without external help

b) Program is divided into individual Scrums based on technical components. Each Scrum Team has its component specific skills needed to finish their own component without external help

c) Program is organized into consumer Scrums and service provider Scrums (front end could be consumer who plays as Product Owner to a middle tier Scrum). Each Scrum gets the dependencies work done leveraging their Product Owner position

66. How are the Non-Functional Requirements addressed by the Scrum Team?

a) by testing them in 'Hardening Sprint'

b) by ensuring that they are met by every Increment and typically defining them in the Definition of Done

c) By having a Non-Functional System Team owning them

67. A Scrum Team has technical specialists in its composition. The specialists perform their work when the Sprint Backlog needs their special skills, but they are idle otherwise.

a) Continue to have the specialists to deliver fully integrated Increments. Gradually facilitate the Developers to organize their work to fully leverage these special skills. If required, the team can enhance everybody's domain of expertise, so everyone is productive as team without idle time

b) Let the project manager coordinate their staffing needs and plan partial allocations to different teams to avoid idle time

c) Defer and accumulate the special work to later Sprints until it needs full time specialists. Deliver the Increment with workarounds. Later, when specialists are added, refactor the Increment removing the workarounds so it can become releasable.

68. The Product Owner provides the transparency of their product plan to the stakeholders and the Scrum Team through

a) Planning Backlog
b) Sprint Backlog
c) Project Backlog
d) Product Backlog

69. A Scrum Team needs to develop a web application in Increments. Some of the Sprints have Sprint Goals like this: 'Develop Data layer for Functionality A'. What is your inference?

a) The Scrum Team follows horizontal decomposition of the Product Backlog Items. This is recommended
b) The Scrum Team follows vertical decomposition of the Product Backlog Items. This is recommended
c) The Scrum Team follows horizontal decomposition of the Product Backlog Items. This is NOT recommended
d) The Scrum Team follows vertical decomposition of the Product Backlog Items. This is NOT recommended

70. The Developers can deliver an Increment that meets the Definition of Done, but the Increment still can have defects that are known to the team and the Product Owner.

a) Yes
b) No

71. Select all that apply. During the Daily Scrum, the Scrum Master's role is to:

a) Facilitate discussions of the Developers
b) Moderate and control so that everyone gets a fair chance to speak
c) Ensure that all 3 questions have been answered
d) Teach the Developers to keep the Daily Scrum within the 15 minute time box
e) All of the above

72. For the Product Backlog Refinement act, the Scrum Team needs to define a recurring pre-set time every week outside the current working hours of the Developers.

a) True
b) False

73. Burn-up and Burn-down charts show evolution of progress over time. In particular,

a) Burn-up shows increase in completion, while Burn-down shows remaining effort
b) Burn-up shows increase in team productivity, while Burn-down shows decrease in productivity
c) Burn-up shows increase in turn-around time, while Burn-down shows decrease in turn-around time

74. The Developers meet every day to inspect the progress and adapt the next day plan. If the Daily Scrum exposes the need to re-plan rest of the Sprint, these re-planning activities happen

a) During the Daily Scrum
b) Immediately after the Daily Scrum
c) As soon as the team gets some extra time
d) The Sprint plan cannot be revised except during Sprint Planning

75. A Scrum Team decides to have an exclusive Sprint to evolve the technical architecture. The sole outcome of this Sprint is a finalized architecture design.

a) It is a good practice since it will help the design to emerge

b) It is not the Scrum approach, since every Sprint must produce at least one releasable functionality

c) It does not matter, since the team is self-managing about how to perform their work

76. In Scrum based software development effort, while the Sprint Goal will deliver a Product Increment, one of the Product Backlog Items is asking for production of a document.

a) It is not okay. Every Product Backlog item must be about a working software requirement

b) It is not okay. Documentation is not needed until Product Owner chooses to release an Increment to production

c) It is okay. A Sprint can produce a document as a sole outcome of the Sprint

d) It is okay. A Sprint can produce other deliverables like document requested by Product Owner along with working Increment

77. An Organization needs to structure hundreds of Developers into Scrum Teams. You as a Scrum Master will

a) work with the organization management and prepare the best structure for each Scrum Team based on the seniority and skills of the Developers

b) identify required number of Scrum Masters and require them to choose their Scrum Teams

c) facilitate the awareness of the Developers about the goals and objectives of the product development, coach them about Scrum, and let them work among themselves to form the Scrum Teams

78. Select all that apply. It is essential for the Product Owner to have these skills. Usually Scrum Master serves the Product Owner by coaching them

a) Software application development

b) Understanding and practicing agility

c) Coaching team

d) Product planning in empirical environments

79. An organization is on its path to adopt Scrum as its approach to software development. It decides to convert all Project Managers into Scrum Masters.

a) It is good strategy. The project managers already know how to run projects. They just need training on Scrum

b) It will create resentment to project managers, because they will have a small team to manage

c) The organization needs to rethink on this strategy. Identifying persons who are inclined or experienced in coaching and facilitation as their leadership style is a better strategy.

80. Select all that apply. A Product Owner requests the Developers to help them with some tasks related to Product Backlog maintenance.

a) The Scrum Master should step in and coach Product Owner to perform their job themselves

b) It is okay, but Product Owner is still accountable for the Product Backlog maintenance.

c) The Developers should refer Product Owner to speak with their manager

d) The Developers can volunteer if this additional task does not impact their Sprint work

Model Assessment - Answers

1. Correct answer is 'd.' Definition of Done provides the common understanding to the Scrum Team about how to assess the completion of a Product Backlog item or the Increment.

2. Correct answer is 'a.' Scrum does not prescribe or mandate any tools. There is no role such as Scrum coach.

3. Correct answer is 'b.' The real mark of progress in Scrum is - the delivery of useable product Increment in every Sprint.

4. Correct answer is 'b.' Scrum Master is not a people manager. Scrum does not prescribe any technology. Scrum is container framework within which techniques and technologies can be employed to develop complex products.

5. Correct answer is 'c.' Other than the "Sprint," all other four events facilitate inspection and adaptation. Backlog Refinement is called as an Act within Scrum.

6. Correct answer is 'a.' Sprint Review is a Scrum event that offers an opportunity to inspect and adapt. Stakeholders collaborate to review the timeline, budget, potential capabilities, and marketplace for the next anticipated release of the product. The team also explains what happened during the Sprint. But they do not inspect about the defect and causes.

7. Correct answer is 'a.' The Scrum framework is a collaboration framework within which Scrum Team can creatively and productively deliver business value with quality. The team model in Scrum is designed to optimize flexibility, creativity, and productivity.

8. Correct answer is 'b.' Sprints are done consecutively, without intermediate gaps.

9. Correct answer is 'e.' Sprint Planning is focused on coming up with Sprint Backlog and Sprint Goal. Sprint Backlog consists of work planned for that Sprint, the plan to achieve that work, and one or more team improvement items. Technical architecture is evolved over the Sprints.

10. Correct answer is 'b.' Scrum allows the Product Owner to decide what the team will work on next by ordering the Product Backlog Items. In Sprint Planning, the team picks up these backlog items as the scope of the Sprint. However, after the Sprint Planning and until the Sprint end, the Product Owner cannot correct the Sprint Backlog without the Developers' consent.

11. Correct answer is 'c.' The Scrum events are strictly time boxed. They end as per the time box no matter what.

12. Correct answer is 'c.' Retrospective is an event where the team inspects their way of working (people, relationships, processes, and tools), and adapts any improvements.

13. Correct answer is 'b.' Scrum does not mandate techniques like Sprint burn down or earned value. However, it stresses bringing-in highest transparency of the underlying information behind Scrum artifacts.

14. Correct answer is 'b.' While the Product Owner has the final say on the content and order of the Product Backlog, he can still get the input / recommendations / ideas about new items from any stakeholders for consideration.

15. Correct answer is 'b.' Every item in Product Backlog needs to have a description, order, value, and estimate. The Product Owner works with the Developers throughout in Backlog Refinement sessions, to refine the backlog items and get the estimate.

16. Correct answer is 'c.' Sprint Review is a Scrum event to inspect and adapt the product development. Sprint Retrospective focuses on inspecting and adapting the way of working to develop the product.

17. Correct answer is 'b.' All these issues have something to do with collaborating with business stakeholders, maintaining Product Backlog, participating in Scrum events, etc. These are Product Owner's responsibilities.

18. Correct answer is 'c.' There are many tools like product burn-down that help to show the evolution of the past and its projection into future. While they are useful, none of these tools are mandated by Scrum. Scrum Master should strive to coach the team about importance of empiricism and not the tools.

19. Correct answer is 'a.' Sprint Goal is the purpose of the Sprint and hence needs to be preserved. The Sprint Goal gives the Developers some flexibility regarding the functionality implemented within the Sprint.

20. Correct answer is 'd.' Scrum does not prescribe any specific estimation technique.

21. Correct answer is 'b.' Every Sprint includes producing a valuable and useful Increment. However, it is Product Owner's call to release that to production.

22. Correct answer is 'c.' In topic three, the Developers put together a plan of how to achieve the scope of the Sprint. It primarily involves deriving work tasks. Since they are accountable to complete these tasks, they are more active during topic three.

23. Correct answer is 'b.'

24. Correct answer is 'c.' Scrum will expose all weakness in the current eco-system that need to be acknowledged and resolved by the organization.

25. Correct answer is 'b.' Though stakeholder is generally regarded as those having some interest in the product, Scrum has this specific definition of the stakeholder.

26. Correct answer is 'c.' Implementation of Scrum does not require any tool.

27. Correct answer is 'a.' The testing approach is part of development work. The development work is owned by the Developers.

28. Correct answers are 'a,' 'c,' and 'd.' Retrospective is an opportunity for the Scrum Team to inspect and adapt the Scrum Team itself.

29. Correct answer is 'c.' The Developers own the Sprint Backlog.

30. Correct answers are 'b' and 'd.' Sprint Review is an informal meeting, not a status meeting, and the presentation of the Increment is intended to elicit feedback and foster collaboration. There are no sign-offs.

31. Correct answer is 'a.' Multiple teams working on the **same** product must have a **single** Product Backlog. It is the responsibility of all the teams to mutually define their Definitions of Done, and then work with each other so that they can create an integrated Increment that is valuable and useful.

32. Correct answer is 'a.' The Sprint estimation is not necessarily constant. As more is learned, the work is adjusted. When new work is added, it increases the amount of remaining work.

33. Correct answer is 'c.' Every Increment is a valuable and useful Increment. It means that whatever is required for the release, it should be defined as part of Definition of Done. For a Product Backlog item to be considered as 'complete', it must meet this Definition of Done.

34. Correct answers are 'a,' 'c,' and 'e.' Other answers also reflect Scrum. However, these five are the Scrum Values as defined in The Scrum Guide.

35. Correct answer is 'c.' Since everyone is working on a different product, there is minimal chance of teamwork, collaboration, and team self-management.

36. Correct answer is 'a.'

37. Correct answer is 'a.' Scrum is lightweight with Scrum Teams and their associated simple roles, events, artifacts, and rules. But for new teams, it is difficult to master the skill to produce deployable and useable Increment within short Sprints. Scrum is not associated with any tools.

38. Correct answer is 'd.' Scrum events are time boxed. Sprint needs to be over by defined date. However, the scope of the Sprint may expand or contract as more is learned throughout the Sprint. When new issues emerge that threaten the completion of Sprint by pre-set date- As a first step, the team needs to capture this as an issue and try to solve on their own. If they cannot, they should make this impediment transparent and take Scrum Master's help. Even after that, if the impediment is not solved, they need to involve the Product Owner to discuss the alternatives.

39. Correct answer is 'b.' It is responsibility of the Product Owner to track the progress of the Product Backlog and forecasts the completion. This forecasting is done in every Sprint Review.

40. Correct answer is 'c.' The first step is to self-manage the issue and find work-around to preserve the Sprint Goal completion. If that does not solve the issue, it needs to be raised as an impediment seeking Scrum Master's help.

41. Correct answer is 'a.' Sprint Backlog contains the Product Backlog Items for the current Sprint, and the plan to complete and realize the Sprint Goal. Scrum does not prescribe any specific format or technique to be followed for representing Sprint Backlog.

42. Correct answer is 'c.'

43. Correct answer is 'a.' It is an optional standard, tracked by the Developers for use within the Scrum Team.

44. Correct answer is 'b.' The Scrum Master does not manage people. They encourage the self-management of the team to manage its work. However, the Scrum Master is the guardian of the Scrum framework and hence its rules. The Developers should only work on tasks related to Sprint Goal. When there is a violation, Scrum Master actively steps in to coach the team on Scrum.

45. Correct answer is 'b.' Every Sprint is meant for delivering an Increment of releasable software / product, as required in Sprint Goal. If an outside item is taking the team's time, it is treated as an issue. If the team is forced by any authority, Scrum Master needs to coach these external authorities about how Scrum works and facilitate removing this issue.

46. Correct answer is 'a.' The activities required to complete the Product Backlog Items to a Done state should be completely within the ownership and influence of the Scrum Team.

47. Correct answer is 'a.' A Scrum Team should be cross functional enough, i.e., should have all the skills needed to convert the Product Backlog Items into Done Increment. If the team needs to depend on external entities for converting backlog items into done Increment, it is not cross functional enough. Every Sprint outcome should be valuable and useful. Mocking up does not complete the work as per that standard.

48. Correct answer is 'c.' The Developers should maintain highest transparency while making a forecast of the work that they believe they could complete. In this case, they cannot do that because the Product Backlog Items do not provide enough information. So, they have to utilize the time available to refine the items to required state and proceed with plan. Later, in the Retrospective the Scrum does discuss the root cause and hence answer 'd' is also correct. But the question asks about "next best thing to do."

49. Correct answer is 'a.' Scrum events are time boxed. Sprint needs to be over by defined date. However, the scope of the Sprint may expand or contract as more is learned throughout the Sprint. When the work gets contracted due the new findings, and there is more room for additional work, the team makes it transparent to the Product Owner. Scrum Master mentors the team to increase such transparency.

50. Correct answer is 'c.' The Scrum Master does not manage people. They encourage the self-management of the team to manage its work. However, the Scrum Master is the guardian of the Scrum framework and hence its rules. Daily Scrum is an opportunity to inspect and adapt daily progress, so that the work-related differences are not allowed to go beyond a day. When there is a violation, Scrum Master actively steps in to coach the team on Scrum.

51. Correct answer is 'a.' When new team members join, the productivity of the team will be temporarily reduced.

52. Correct answer is 'b.' Technical debt is not a concept within Scrum. However, it is commonly used by Scrum Teams to indicate the gap unaddressed in a done Increment.

53. Correct answer is 'a.'

54. Correct answer is 'b.' It is the responsibility of the Product Owner to verify that a Sprint has enough value to worth the effort. They are rigorous value optimizers.

55. Correct answer is 'b.' Since all of them work on a single product, there will be one common Product Backlog. But each Scrum Team will have its own Sprint Backlog and Sprint Goal.

56. Correct answer is 'b.' Scrum will expose all weakness in the current eco-system that need to be resolved. Scrum does not define any organization change management process.

57. Correct answer is 'c.' Empiricism is a process control theory in which only the past is accepted as certain and in which decisions are based on observation, experience, and experimentation.

58. Correct answer is 'a.' There is no exclusive Sprint only to finalize the design. Every Sprint must be used to produce at least one working functionality that is valuable and useful.

59. Correct answer is 'b.' The Product Owner is the ultimate authority of the Product Backlog on which the Developers must work. Those wanting to change a Product Backlog item's priority must address the Product Owner. For the Product Owner to succeed, the entire organization must respect his or her decisions. If the Developers are given different work, it indicates that Product Owner's authority is interrupted.

60. Correct answers are 'a' and 'c.'

61. Correct answer is 'c.'

62. Correct answer is 'a.' The Product Backlog is never complete during the project. It undergoes constant changes and continuously refined. It exploits emerging opportunities and adjusts the emerging risks, so the value can be optimized.

63. Correct answer is 'c.' Cancelation of the Sprint is decided by Product Owner, and Product Owner will not cancel the Sprint unless the Sprint Goal becomes obsolete. Here the Sprint Goal is intact. Also, the Sprint duration cannot be extended since it is time boxed.

64. Correct answer is 'a.' Definition of Done is a standard to define the quality measures required for a useful product. Acceptance criteria is the specification of expected business features (functionality) of the product.

65. Correct answer is 'a.' It is preferable to divide teams such that each team has absolute ownership of their work without external dependencies.

66. Correct answer is 'b.' Definition of Done defines the standards to be met for a Product Backlog item to be considered as Done. Typically, Non-Functional Requirements are added to the Definition of Done so that such requirements are built into every Increment.

67. Correct answer is 'a.' There is no project manager role in Scrum and the Developers manage their own development

work. A Scrum Team must be cross functional enough, that is, it should have all required special skills, without the need for any external help in completing the Sprint Backlog.

68. Correct answer is 'd.' The Product Owner uses Product Backlog to update the stakeholders on the current state of the product plan.

69. Correct answer is 'c.' It is preferable to decompose the Product Backlog Items such that each team can produce fully working business functionality on its own rather than producing a technical component. Such decomposition is called as vertical decomposition. A horizontal decomposition on the other hand makes the team depend on other teams to integrate and create a fully working business functionality. Such teams usually end up as just technical component teams.

70. Correct answer is 'a.' An Increment can have known gaps but must meet the Definition of Done. The reason for having a lenient Definition of Done is - Definition of Done should contain conditions that are realistic to achieve for the team. For newer teams, they can start with a realistic (feasible for the team) Definition of Done, and then can be continually be improved by maturing team's ability to perform all that is required to deliver flawless Product Increment. Having a realistic Definition of Done for newer team means that the working Increment may have known bugs. But such gaps are transparent between the Developers and Product Owner.

71. Correct answers are 'a' and 'd.' Scrum Master facilitates the Scrum events as and when requested by others or required by their observations. Scrum Master does not take any active role in directing or controlling the Daily Scrum. It is up to the Developers to fully leverage it for their synchronization and progress. Scrum Master is the guardian the Scrum process and time boxing is a cardinal rule of Scrum. So, Scrum Master coaches the team to keep the Scrum rules.

72. Correct answer is 'b.' This is an ongoing act that happens within the hours of current Sprint. The time can be mutually discussed and agreed by Product Owner and the Developers. Usually it does not take more than 10% capacity of the Developers. Also, Product Backlog Items can be updated at any time by the Product Owner or at the Product Owner's discretion.

73. Correct answer is 'a.' Both burn-up and burn-down are not mandatory but optional in Scrum. They are used to make the progress transparent.

74. Correct answer is 'b.' The Developers use the Daily Scrum to inspect progress towards the Sprint Goal and to inspect how progress is trending towards completing the work in the Sprint Backlog. During the Daily Scrum they come up with the next 24 hour plan. But, if they see that the entire Sprint plan needs a revisit, they meet immediately after the Daily Scrum for detailed discussions, or to adapt, or re-plan, the rest of the Sprint's work.

75. Correct answer is 'b.' In Scrum, technical architecture is evolved continuously throughout the project, as more is learned. There is no exclusive Sprint or Scrum event to define the technical architecture upfront. Usually the Developers define architectural guidelines that every team member can use in their work. The team can also have core hours to review the design and architecture during the Sprint. As a result, these guidelines are continually updated, and the technical design emerges on the go.

76. Correct answer is 'd.' While the Sprint must necessarily produce a valuable and useful Increment, some of the Product Backlog Items could produce other deliverables including documents if the Product Owner considers them having appropriate value. If a Product Backlog item is a document, it may not be subjected to Definition of Done which is usually the standards needed for software. So, while the Definition of Done is applicable at Increment level, it may not be applicable for some individual Product Backlog Items.

77. Correct answer is 'c.' Scrum Teams are self-managing teams. Given the knowledge of the product vision and sound understanding of how Scrum works, the team is knowledgeable enough to form themselves into Scrum Teams. A Scrum Master needs to facilitate this.

78. Correct answers are 'b' and 'd.' Product Owner must have the understanding to perform product planning in empirical environment, and practicing agility. Scrum Master serves the Product Owner by coaching them these skills.

79. Correct answer is 'c.' Scrum Master does not manage any team. The Scrum manager is not required to know project management since it is shared between three roles of Scrum.

80. Correct answers are 'b' and 'd.' The Product Owner may have the Developers help them with Product Backlog maintenance. However, the Product Owner remains accountable. There is no manager for the Developers within the Scrum Team.

#

Journey to Excellence is a Path. Good Luck on Your Journey!

About the Author

Mohammed Musthafa Soukath Ali

SCJP, LOMA 286, PMP, PSM, PSPO, SA, SPC

Musthafa led one of the largest Agile transformations in corporate history, 'Tata Consultancy Services (TCS) Enterprise Agile by 2020.' He is the author of the #1 Best Selling book Scrum Narrative and PSM™ Exam Guide on Amazon and smashwords.com. He is the architect behind the prestigious TCS Agile Ninja Coach program and has trained over 500 Agile Coaches. He is the co-inventor of the TCS Location Independent Agile™ method and the TCS AgilityDebt™ framework. He is also the external Management Capability Adviser for some of the Tata Group Companies. He is ranked as one of the TCS Global Top Project Planners and Agile delivery experts, having consulted with 15+ global customers. He published 9 papers in conferences.

You can connect with the author through his LinkedIn network: https://in.linkedin.com/in/mohammed-musthafa-soukath-ali-9a857762

About the Editor-in-Chief

Samantha Mason

PSM, CSM

Samantha began as a software engineer at Altsys Corporation, later acquired by Macromedia, working on the graphics program FreeHand. Her teams worked in an Agile way before the term was coined. Her expertise in real-world development provides a unique advantage in editing and presenting information aimed at those new to Agile developments. Samantha is also the editor of the #1 Best Selling book Scrum Narrative and PSM Exam Guide in smashwords.com. Being a trained facilitator for The Knowledge Academy, she has access to students new to Scrum and Agile and their interest in SAFe, which allows her to continuously update publications to address ambiguous areas. Currently she works out of Mequon, Wisconsin.

You can connect with the editor through her LinkedIn network:

https://in.linkedin.com/in/samantha-mason-956b069b

##

Made in the USA
Coppell, TX
02 September 2022

82536026R00099